Mathematics Achievement of Immigrant Students

Dirk Hastedt

Mathematics Achievement of Immigrant Students

 Springer

Dirk Hastedt
IEA
Amsterdam, The Netherlands

ISBN 978-3-319-29310-3   ISBN 978-3-319-29311-0   (eBook)
DOI 10.1007/978-3-319-29311-0

Library of Congress Control Number: 2016931290

Printed on acid-free paper

This Springer imprint is published by Springer Nature
The registered company is Springer International Publishing AG Switzerland

Preface

This book is a modified version of a dissertation prepared at the University of Vienna for a PhD in education. When I started working on this undertaking, I aimed to examine changes of populations in international large-scale assessments where the populations of countries are usually assumed to be constant in most research. This topic was brought to my attention by Hans Wagemaker, to whom I am indebted not only for this inspiration but also for his encouragement while I was working on the dissertation.

In view of current events in Europe and the Middle East, the topic of immigrants, and immigrants in education systems, has provoked keen interest and gained relevance – especially in these regions. I hope that this publication helps to better understand the situation of immigrant students and that it inspires researchers as well as policy makers to improve the situation of immigrants and education systems that face severe challenges due to high immigration rates. The examples of Singapore and Canada show that improving the situation for immigrants in schools is not only a matter of education policies. It also seems to be influenced by the opinions toward immigrants and by policies fostering positive attitudes toward immigrants in general.

Some of the results presented, such as the differences between immigrant boys and girls, will hopefully encourage some study participants, e.g., Eastern European countries where the differences are particularly pronounced, to analyze this aspect more in depth. The importance of school safety – especially for immigrant students – will hopefully also be acknowledged. I expect that the results regarding the peer effect might change the current perspective that the presence of immigrant students in class always has a negative impact on their peers' achievement. Supporting immigrant students is also an indirect means to help their native peers because high-achieving immigrant students seem to have a positive effect on their peers' achievement.

I would like to conclude these remarks by thanking all the people who helped me conduct this research, and I would like to mention a few of them who played a very important role.

I would like to thank Stefan Hopmann for serving as my doctoral father and for his manifold support during the preparation of this dissertation. My thanks also go to Henning Schluß and Tamara Katschnig and many others at the University of Vienna for their continuous help.

I am deeply grateful for the feedback and support I got from Seamus Hegarty and David Rutkowski that helped improve this publication tremendously. The task of layout and editing was done wonderfully by Viola Hingst – thanks!

Finally, my deepest thanks goes to my wife Maren and my kids Helen, Hannah, and Emily who tolerated me being stressed when writing this dissertation and who always encouraged me to continue working on it. Moreover, my mother's trust in me was of tremendous value throughout my whole life but especially during the busy times of writing this piece of research.

Amsterdam, The Netherlands   Dirk Hastedt

Contents

About the Author

Dirk Hastedt has over 25 years of experience in the field of international large-scale assessments. He joined the International Association for the Evaluation of Educational Achievement (IEA) in 1989 (then the IEA Germany foundation) and helped shape the IEA Data Processing and Research Center (DPC) since its establishment in 1995. As a Senior Researcher at the IEA DPC, he was Project Manager for data processing for IEA's Trends in International Mathematics and Science Study (TIMSS), and he also headed the data processing for various IEA studies, such as the Progress in International Reading Literacy Study (PIRLS) and the International Civic and Citizenship Education Study (ICCS). In 2001, he became Co-Director of the IEA DPC, where he was responsible for the center's international field of work and was involved in initiating pioneering undertakings such as the IEA International Computer and Information Literacy Study (ICILS). As of 2014, Dr. Hastedt is IEA's Executive Director, overseeing the organization's operations, studies, and services and driving the overall strategic vision of IEA.

Dirk Hastedt holds a PhD in education from the University of Vienna and a diploma degree in mathematics from the University of Hamburg. His research interests focus on quantitative research, statistics, survey research, mathematics, and policy analysis.

He serves as board member of the IEA-ETS Research Institute and as coeditor in chief of the IEA-ETS Research Institute's journal *Large-scale Assessments in Education*.

Chapter 1
Introduction

Abstract An increase of immigrants and consequently students with an immigrant background is one aspect of the globalization that takes place in our world today. This chapter summarizes the discussion about globalization with an emphasis on immigrants and immigrant students. Immigrant students are a quite diverse group of students – some might have highly skilled parents who are looking for job opportunities in another country, whereas others might be war refugees. Different groups of immigrant students have different needs, which must be taken into account when it comes to their education. The countries thus face different challenges in catering to these needs and in providing a good education for students with an immigrant background. Offering a high-quality education to all students is also a question of social justice.

Keywords Globalization • Immigrants • Social justice

This publication examines the mathematics achievement of immigrant students in several countries. An increase of immigrants and consequently students with an immigrant background is one aspect of the globalization that takes place in our world today. Immigrant students are a quite diverse group of students – some might have highly skilled parents who are looking for job opportunities in another country, whereas others might be war refugees.

Different countries face different challenges in catering to and providing a good education for students with an immigrant background. Offering a high-quality education to all students is also a question of social justice.

This publication aims to contribute to the discussion on immigrant students in educational systems and to help understand differences between and within countries. It will especially explore circumstances and policies that might lead to a better education of immigrant students. This work seeks to make recommendations for policies that promote a better education for students with an immigrant background. For this purpose, quantitative research methods will be used and IEA TIMSS data will be analyzed. In a second step, in-depth policy studies will be conducted for countries that achieved positive results in terms of the immigrant students' education in the quantitative analysis of TIMSS data. With this approach, the aim of this

© Springer International Publishing Switzerland 2016
D. Hastedt, *Mathematics Achievement of Immigrant Students*,
DOI 10.1007/978-3-319-29311-0_1

publication is also to demonstrate how quantitative analyses of large-scale assessment data can be augmented with more in-depth policy studies.

The publication first gives a general introduction into the topic followed by a review on the current state of research in the areas addressed. A description of the data used and the methods applied will be made. To follow are the results of the analysis of TIMSS data. The analysis will at first keep a broader focus on all cycles of TIMSS until 2007, examining the percentages of immigrants and their achievement before concentrating on the 2007 data for more in-depth results. Particular attention will be paid to students' backgrounds to better understand how the immigrant student population differs between and with the countries.

Differences between schools and classes will be examined, and in the following two chapters, policy studies will be presented on two countries that came out as successful in the quantitative analysis of TIMSS data. Finally a concluding chapter will bring together and discuss the different results.

Globalization

Today's world is strongly influenced by globalization. But what is globalization? The term globalization is used widely but quite divergently and often very vaguely. Mundy (2005) even called the term "muddy." She defines globalization as a combination of four effects, which are the mobility and territorial spread of production chains, the concentration of production into the hands of large, multi- or transnational corporations, the financial and capital flow crossing country borders, and the emerging of the new information economy.

Similarly, Blossfeld et al. (2008) understand the process of globalization as an interaction of four structural developments. The first one is an increase of the international exchange of goods, services, and labor: the international trade has increased, and workers migrate between countries and regions. The second development is an increase in the business competition between countries and regions. The boundaries and restrictions of the markets were loosened and international companies open and close branches worldwide depending on what is offered by the countries and the prices for production. The third development is fostered by the developments of modern technologies – especially information technology. The Internet and other modern information technologies enable people to cooperate and to exchange information in a formerly unknown speed. The fourth development concerns the internationally linked markets and as a consequence instability and vulnerability of local markets.

All these definitions focus mainly on the economic aspects of globalization. Burbules and Torres (1999) distinguish different terms of globalization. They see globalization in economic but also in political and cultural terms. In political terms, globalization is the loss of the sovereignty of the nation states. In cultural terms, they describe the tension between more cultural homogeneity between countries on one side and more fragmentation or heterogeneity on the other. They see all these three aspects influencing the education systems of the countries in the world (p. 14).

Although Blossfeld et al. (2008) define the term globalization as described above, they also use the term differently later when addressing the globalization of school systems. In the context of globalization of school systems, they use the term globalization as the development of a common curriculum across countries or as "importing" a curriculum from another country, which the authors conclude is not new but goes back to the roman term of "septem artes liberals" which influenced the education within Europe up until the middle ages (p. 67).

This publication will focus on the migration with a special emphasis on the effect on the school systems around the world. Migration poses a challenge to countries, and an increasing migration between countries and regions also affects their educational systems. The educational systems have to cater to students with different cultural backgrounds, different mother tongues, and more diverse socioeconomic backgrounds, which can be observed in a growing number of countries (Castles, 2009). The increased mobility of employees and the growing demand for work forces in various regions of the world generate a steady flow of immigrants and their families (OECD, 2012a). But economic reasons and job seeking are only two of the reasons for migration, and thus the term "immigrant" is used for a heterogeneous group. Wars, civil disorder, natural disasters, and climate changes also create a significant flow of immigrants, and so the number of people – and consequently also children – affected is huge.

The United Nations reported: "In 2010, the Office of the United Nations High Commissioner for Refugees (UNHCR) counted 43.3 million forcibly displaced people worldwide – the highest number since the mid-1990s. This included 27.1 million internally displaced persons (IDPs), 15.2 million refugees and 983,000 asylum-seekers. Of the 15.2 million refugees, 10.4 million were under UNHCR's responsibility, and 4.7 million were Palestinian refugees under the mandate of the United Nations Relief and Works Agency for Palestine Refugees in the Near East" (United Nations, 2014).

These immigrants especially, and in particular the children in this group, require special attention. There is a wide range of research on mental and psychological problems of refugee children. It is clear that this population is very vulnerable and faces particular challenges, e.g., some refugee children need special attention to overcome their traumatic experiences, and others are separated from friends and family.

However there are more differences. Some immigrants are highly qualified and educated, while others are less so; some work in highly paid jobs and others in poorly paid jobs. This is, for example, reflected in the cultural types that Baumann (2003) specified as "tourists" and "vagabonds" where "… the tourists are the privileged 'free consumers', whereas the vagabonds are the underprivileged 'flawed consumers'" (Jacobsen & Poder, 2008, p. 144). Different immigrants have also different perspectives: some want to become permanent residents while others want, or have, to leave their country of residence after a certain period of time (OECD, 2010a).

Educational systems are influenced by globalization in several ways. One is the increased number of students with an immigrant background for which educational systems have to cater. Another is the influence of international and transnational organizations such as the United Nations, the European Commission, the World Bank, or the OECD.

Transnational organizations such as the European Commission are setting standards for educational systems that countries are trying to achieve. Loans from the World Bank or the Inter-American Development Bank are linking loans to the achievement of educational outcome measures, which impact the educational systems. Moreover the increased economic competition between countries creates competition among the educational systems. International companies are moving the production between countries, basing their decisions on economic considerations as explained in the following.

Green argues: "Where regional economies come to be dominated by transnational corporations which have no national allegiance and which are beyond national government control, there can be no such thing as a national economy" (Green, 1997, p. 154). Part of these considerations is also the availability of adequately educated work force.

As an example for this, I refer to the policies of the city of Chicago in the United States. Rahm Emanuel, the mayor of Chicago, explained the changes in the educational system of Chicago at a World Bank symposium. He mentioned that the city had recently changed its educational policies and linked education more closely to the needs of big companies on the one hand to make the city of Chicago more attractive for the industry and on the other hand to improve the perspectives of graduates from schools in Chicago by increasing their chances on the labor market. Companies are represented in the school boards of some charter schools in Chicago to ensure that the curriculum of the schools matches the job requirements in their companies. As a result, some big companies have moved their headquarters to Chicago. Moreover the graduation rates at charter schools and the percentage of graduates who find jobs after graduation have increased (Emanuel, 2013).

Although this example can be regarded as a success of the economic and maybe also the educational system of a city, it is also an example of the economic competition of regions – in this case within a country – and the impact it can have on educational systems. An educational system that creates a labor force needed by the industry has an economic advantage because it attracts the industrial production. Green (1997, p. 176) also describes: "In most countries there have been attempts to reinforce the institutional linkage between education and work, through the development of work-experience, work shadowing, alternance and mentoring programs. Employers have been drawn more systematically into the process of standard-setting under the Commissons Consultatives Professionelles (CPCs) in France, the BIBB in Germany and the Industry Lead Bodies in the UK."

The globalization process increased significantly in recent years and changed the challenges faced by educational systems around the world (Green, 1997; Rizvi & Lingard, 2010).

Since the 1970s, the world has undergone some changes since globalization affected not only the economics of countries but also their societies and their educational systems in particular (Green, 1997). Green argues that "... the dynamics of education within the older nation states have changed. As nationhood has been consolidated and sustained, and with the growing international economic competition in the postwar period, education has partially lost sight of this formative mission and purpose, in the advanced states now, with the possible exception of Japan, education is seen primarily as a means of individual and collective economic advancement" (p. 4).

One of the most influential organizations on developed countries' educational systems is the OECD, which develops indicators setting standards for educational systems (Lingard & Grek, 2007). The OECD was originally funded by the United States under the Marshal plan to rebuild the European economy. In the mid-1980s the OECD launched its project on educational indicators (INES) mainly under the influence of the United States (Henry, Lingard, Rizvi, & Taylor 2001). The collection of key indicators in education and the potential of reducing educational systems to few simple measures were found problematic even before it started and even within the OECD as Heyneman (1993) described: "Those [from CERY staff – addition by the author] whom I interviewed believed it was unprofessional to try and quantify such indicators, and that it would oversimplify and misrepresent OECD systems, ..."

Rizvi and Lingard (2006, p. 133) argue that "... the OECD is... promoting a policy agenda for reforming educational governance, based on neo-liberal precepts of marketization and privatization on one hand and strong systems of accountability on the other" and that "It highlights the relevance of these governance principles for all its member countries, regardless of their local histories and traditions..."

Rizvi and Lingard's conclusions on the neoliberalism of the OECD, however, were criticized as being quite counterfactual reasoning and leaving out the globalization process itself (Lauder, Brown, Dillabough, & Halsey 2006, Chapter 17).

One impact of this globalization process on educational systems that some researchers found was a convergence of educational systems. For example, Blossfeld et al. (2008) argue that educational systems are becoming much more alike. They concluded: "... seit den 1990er Jahren durchgeführte international Schulleistungsuntersuchungen wie TIMSS, PISA und PIRLS/IGLU dokumentieren universelle Kongruenzen der Bildungssysteme, die durch Prozesse der Internationalisierung und Globalisierung gefördert werden" ("... international assessments of student achievement conducted since the 1990s such as TIMSS, PISA, and PIRLS/IGLU document universal congruencies of the education systems which are fostered by processes of internationalization and globalization"[1]). We can see that curricula around the world are becoming much more similar. This has, on the one hand, advantages for students migrating from one country to another because the subjects taught are more similar and a transfer potentially easier. On the

---

[1] All translations from German are my own.

other hand, it poses the question if these curricula are adequate for the students in the countries.

For the developing countries, the history of external influences on the educational policies goes even back further than the 1990s, i.e., to the middle of the last century and to the influence of the World Bank "... through accountability demands of structural adjustments attached loans and aid" (Lingard & Grek, 2007, p. 2). Loans and aids were – and still are – connected to the implementation of educational reforms. Some of these reforms had positive effects on education. More financial resources were allocated to education, more teachers were employed, more and better-equipped classrooms were built, and more learning resources were made available to the students. Other changes, however, even had negative impacts on the countries' education. The Education For All initiative increased the percentage of students included in the education systems of several developing countries. This was especially beneficial for girls who were left out to a greater extent than boys in some countries. But some developing countries did not have sufficient resources to deal with such an increase in the student population. For example in Uganda, school fees were abandoned in 1996. This resulted in an increase of children attending schools from 2.7 million pupils in 1996 to 5.3 million pupils in 1997. Since it was not possible to build new schools and classes as well as employ additional teachers to the extent needed, the teacher-pupil ratio increased from 1:49 to 1:120 (in 1998, additional teachers were employed, which resulted in a decrease of the pupil-teacher ratio to 1:80) (Tappy, 2008).

These examples show that not all changes imposed by external pressure on educational systems – especially of the developing countries – lead to positive results. The international organizations learned from these experiences and the focus has moved from Education For All to Learning For All. The indicators that were developed in the Education For All agenda that focus on participation in primary education are in the process of being changed to more outcome-oriented indicators in the post-2015 agenda (UNESCO, 2013b; United Nations, 2013; World Bank, 2013). What these indicators will finally look like and how they will impact educational systems have yet to be seen.

Coming back to the increased number of immigrant students and their diversity, I note the following consequences: In Europe, several countries' educational systems face major difficulties educating immigrant students or students with different cultural backgrounds and mother tongues, and they have taken various measures to overcome these problems (Eurydice network, 2009). Additionally, there are also traditional immigration countries like the United States, Canada, or Australia where language problems and different backgrounds and cultures of people existed from the beginning of the foundation of their educational systems. Indeed, Kennedy (2012, p. 14) argues that "In new nations, such as the USA, education also had played a major part in assimilating immigrant cultures." But does their experience make them better at dealing with these problems?

Globalization and immigration and the effect that they have on educational systems are global challenges of today's world, and it is vital that we gain a better understanding of the topic and its magnitude – and find solutions that may help policy makers in managing these challenges.

What Matters to Immigrant Students

Students with an immigration background face several challenges in their school career. The most obvious one is that students with an immigrant background might be challenged by not being fluent in the language of instruction. Buchmann and Parrado (2006), for example, found that language difficulties are often considered as the main factors for the lower performance of immigrant students. However, also cultural differences between their country of residence and the culture experienced at home or a low socioeconomic background can disadvantage these students. Schwippert et al. (2007, p. 63) found: "Ein nicht unerheblicher Teil des Leistungsrückstandes von Schülern aus Familien mit Migrationshintergrund kann also durch die soziale Lage der Familien erklärt werden, die in fast allen Staaten im Durchschnitt schlechter ist als in Familien ohne Migrationshintergrund" ("Not a small part of the underachievement of students from families with a migration background can thus be explained by the social situation of the families which is worse than average than in families without migration background in almost all cases").

Also Sirin (2005) found that a big portion of achievement differences can be attributed to the socioeconomic background of students. Another important factor is the distribution of immigrant students across classes. As shown by Netten (2010) for the Netherlands, if immigrant students cluster, this has a significant negative effect on their achievement. But research based on PISA data found this to be more of an effect of the students' language, explaining that "A higher concentration in schools of students who do not speak the test language at home is related to worse outcomes for both non-immigrant and immigrant students" (OECD, 2012b, p. 60).

There is not much research so far focusing on teacher and school factors that influence the achievement of immigrant students. Furthermore regional factors such as immigrant students being more prominent in rural or urban areas are not analyzed at the international level. Research on these factors that can be influenced by educational policy – in contrast to, for example, the socioeconomic background of students – is especially important to help policy makers address this challenge adequately.

Outline

The research in this publication is based on the literature review presented in Chap. 2, and there is quite an amount of research on immigrant students and their particular challenges available. The analysis will be based on the IEA Trends in International Mathematics and Science Study (TIMSS). TIMSS is conducted every

4 years since 1995 and measures mathematics and science achievement of students in grade four and grade eight. TIMSS does not only include good achievement measures based on representative samples of students but also rich background information from students, teachers, and school principals. A growing number of countries are participating in TIMSS. Detailed information about TIMSS can be found in Chap. 3.

The research presented here partially tries to examine if the results found in previous research can be replicated with TIMSS data, but will also go beyond by making use of the background data that is unique in terms of its richness and country coverage. The final aim of the analysis is to find conditions and examples where students with an immigrant background successfully participate in an educational system. Success in education can have a broad range of meanings, and different researchers might define success very differently.

For immigrants, this might cover socio-emotional well-being, integration in the host country, maintenance of their own culture and language or preparedness for their future lives, and participation in the society. Neither the data analyzed here nor the volume of a single publication can achieve to reflect on all perspectives of success. The publication at hand will focus on achievement and explore in depth related motivation and self-concept as noncognitive outcomes. Although this is a limitation, these aspects of school success have an impact on the students' chances in future life and well-being, and achieving an increase in these aspects can be seen as one step towards social justice for immigrant students. The analysis will use the following approach:

The first focus is on trends. Based on the previous citations, one would expect that the percentage of immigrant students in several if not most of the countries would increase. Consequently, I will evaluate how the percentage of first- and second-generation immigrant students enrolled in grades four and eight develops between 1995 and 2007.

As described above, the focus here is also on achievement outcome, and therefore, next it will be evaluated *how the mathematics and science achievement of immigrant students compares to the achievement of native students in the various countries in TIMSS and how it changes over time compared to the changes observed for native students in the countries.*

After analyzing these trend aspects, the analysis will focus on the most recent TIMSS cycle – 2007 – and more in-depth analysis shall be presented. The basic student demographics like age, age at immigration, sex, language spoken at home, parents' education, SES background, and students' attitudes are analyzed.

A special emphasis is put on the students' attitudes. As research has shown (see Chap. 2), students' attitudes are associated with their achievement. Some researchers also regard students' attitudes as an important outcome in itself. Consequently, *the differences between immigrant and nonimmigrant students in TIMSS 2007 in terms of the attitudes towards school in general and mathematics in particular as well as their self-esteem in mathematics are analyzed.*

Since the aim of this publication is to find conditions under which students with an immigrant background achieve relatively well, the next area of research is the school level perspective. Since school conditions vary substantially between urban and rural schools in many countries, the first analysis is on types of community. Basic differences between schools that have shown to be relevant for students' achievement will be analyzed. This includes school attendance, school resources, school climate, and school safety.

Finally, differences at class level shall be analyzed. Characteristics that have shown to be relevant shall be analyzed in terms of achievement differences for immigrant students. The characteristics examined are class size, homework given, and concentration of immigrant students in classes.

After this quantitative research, two countries that have shown somewhat positive results for immigrant students are analyzed from a policy perspective. The aim here is to find out *which policies are leading to positive achievement results for immigrant students in Singapore and in Canada and if these can inform the policies in other countries to improve the achievement of immigrant students.*

Chapter 2
Review

Abstract In this chapter, the current state of knowledge about the different aspects influencing student achievement will be reviewed as a basis for analyses to be conducted later in this publication. The variables include the effects of children's age at various stages in the school system, the effect of immigrant students' age at the time of migration, gender differences, and language difficulties which are often considered as the main factor for the lower performance of immigrant students. Achievement differences were proven to be closely related to the students' socio-economic background and parents' education. Further factors influencing students' success are their attitude towards learning and their aspirations, and if they attend school regularly. This chapter will also explore the effects of school-related factors, such as school safety, class size, and homework, all of which are widely discussed in the literature. This chapter will also explain the peer effects that occur, focusing on how nonnative students influence the performance of their native peers.

Keywords Age effect • Gender effects • SES effects • Language effects • Students' attitudes • School resources • School safety • Class size • Peer effects • Homework

So far, the topics of globalization and the situation of immigrant students were introduced, and the research questions that provide the structure of the publication were developed. In this chapter, the current state of knowledge about the different aspects to be analyzed later will be reviewed. The order follows the sequence that will also be applied in the research section.

The number of studies about what causes achievement differences between students is practically endless. The number of analyses carried out to find the differences between students with an immigrant background and native students is great since the topic seems to be of special interest in recent years.

The OECD has a strong focus on research related to immigrants and immigrant students. Several books have been published by the OECD that make use of PISA data but also of other large-scale assessment data as well as policy notes from countries (OECD, 2006, 2010a, 2012a, b, 2013b). The main conclusion is that: "With some exception, immigrant students, on average, have weaker education outcomes at all levels of education" (OECD, 2010a, p. 7). Secondary analysis based on PISA data about students with an immigrant background from other researchers is also increasing almost on a daily basis. Dronkers, for example, has published several

© Springer International Publishing Switzerland 2016
D. Hastedt, *Mathematics Achievement of Immigrant Students*,
DOI 10.1007/978-3-319-29311-0_2

papers (Dronkers & Kalmijin, 2013; Dronkers & Kornder, 2013; Heus, Dronkers, & Levels, 2008). One of his foci in research is to include the country of origin of immigrant students. In his analysis, he found: "Migrant students originating from non-Islamic Asian countries experience higher educational achievement than equivalent migrant children who originate from other countries" (Dronkers, van der Velden, & Dunne, 2012, p. 30), and: "Migrant students originating from Islamic countries experience lower educational achievement than equivalent migrant children originating from other countries" (2012, p. 30).

Effects of Age and Schooling

The students in the different school systems around the world are of different age in different school years, and the age in the different grades also varies. This is partly caused by different policies for school entry but also by other factors.

While in most countries students start school at the age of 6, there are quite a number of countries with different policies. For example, in Australia, Cyprus, England, Jordan, Malta, New Zealand, Palestine, and Scotland, children start school before the age of 6. But also among this group, the policies regarding the school entry age differ. In Australia, school starts for children in the year that they turn 6, but there is some variation between the different Australian states and territories. In Cyprus and Palestine, children must be 5 years and 8 months old to start school. In England, children start to go to school in the term that follows their fifth birthday. In some other countries, children are enrolled in schools at the age of 7, e.g., in Armenia, Bulgaria, Denmark, El Salvador, Latvia, Malaysia, Mongolia, Romania, and Serbia (see Appendix A of (Mullis et al., 2007)).

There are also various forms of policies for grade repetition practiced in the various countries. UNESCO defines grade repetition as "… students are held in the same grade for an extra year rather than being promoted to a higher grade along with their age peers" (Brophy, 2006, p. 6). UNESCO distinguishes between five different forms of grade repetition. There are three forms of voluntary grade repetition imposed by the students or their families – either because there is no school that offers the necessary grade (mostly in rural areas of developing countries) or because of a reduced learning outcome which could be caused by irregular school attendance in developing countries, or due to language problems of the students. Grade repetition can, however, also be caused by students not passing required exams or by involuntary repetitions initiated by schools due to low student achievement. UNESCO concludes that the different forms of grade repetition have different effects also depending on the development level of the country but mostly have a negative impact on the concerned students – especially in terms of motivation, self-esteem, behavioral problems, and finally alienation from schools.

The OECD (2011c) analyzed the number of grade repetitions based on the PISA 2009 data. It turns out that in Macao-China, Tunisia, and Brazil, more than 40 % of the 15-year-old school students have repeated a grade at least once. On the other hand, this never occurred in Norway, Korea, or Japan. "Across OECD countries, an average of 12 % of students reported that they had repeated a grade at least once" (OECD, 2011b, p. 73). While UNESCO is mainly examining the effects of grade epetition on the students, the OECD is focusing on the costs of grade repetition for the educational system and concluded that "In Belgium, the Netherlands and Spain, the costs [of the grade repetition] is equivalent to 10 % or more of the annual national expenditure on primary- and secondary-school education" (OECD, 2011c, p. 2).

As a result, the average age in TIMSS 2007 grade eight by country varies between 13.7 years in Scotland and 15.8 years in Ghana. Discussions are ongoing on how strongly the maturation of students affects the achievement in different subjects compared to the effect of schooling. For example, Cliffordson and Gustafsson (2010) argue that it is as strong as the schooling effect in Sweden. Yet, it seems to be undisputable that age affects student achievement independent of the schooling effect. In the analysis, special attention must be paid to the age of immigrant students compared to their native peers.

Not only the age of students is important; there are also discussions about the effect of the age of the immigrant students at the time of migration. A widespread hypothesis says that the older the immigrants are at the time of migration, the bigger is the obstacle in education. Researchers tried to find a critical age that identifies students who suffer severely from the immigration.

Heath and Kilpi-Jakonen (2012) analyzed this phenomenon by using the PISA 2000, 2003, and 2006 data. They defined "early arrivers" as students who immigrated before the age of 5 and "late arrivers" as immigrant students who immigrated after the age of 12. They found: "Test scores are typically lower for young people who arrive later in their school careers. In other words there are typically 'late-arrival penalties' for the first-generation students, although I also found some examples of 'late arrival premia'. The size of the penalty is much larger for late arrivers than for mid-arrivers (relative to early arrivers)" (p. 27). The OECD (2013b, p. 80) also reported that "immigrant students benefit from an early arrival."

However, there are difficulties with PISA data when analyzing achievement differences because PISA samples 15-year-old students, and students at the age of 15 attend different grades. The analysis of the average age of immigrant students and a comparison to the age of native students revealed that immigrant students attending the same grade are older than their native peers in a number of countries. This means that at the age of 15, immigrant students attend lower grades than 15-year-old native students in a number of countries and consequently have less opportunity to learn. It follows that immigrant students are disadvantaged in the PISA sample which enforces the lower achievement of immigrant students in PISA. Analyses of the PISA data that compare the achievement of immigrant students and native students and do not account for the grade differences tend to overestimate the achievement gap between immigrant students and native students.

Myers, Gao, and Emeka used the US 2000 census data. They used a logistic regression analysis and tried different models for different groups of immigrants depending on their country of origin and distinguished between Latino and Asian immigrants. They examined educational attainment, English language proficiency, and occupation. "The tentative overall conclusion is that young arrival is important to later success, but it helps to know exactly how young. And the effect of young arrival varies substantially across outcomes and groups" (Myers, Gao, & Emeka, 2009, p. 6). Interestingly, they found in their models: "As before, adding a dichotomous variable for arrival under the age of 10 adds significantly to the model" (p. 8) which suggests that there is a critical age of migration that affects the outcome measures.

According to the OECD (2010c, p. 75) "in general, first-generation students who arrived in the host country at a younger age outperform those who arrived when they were older. On average across OECD countries, first-generation students who arrived when they were 5-years-old or younger score 42 points higher than first-generation students that arrived after they were 12-years-old. The size of the gaps, however, varies considerably across countries and across groups." On the other hand, Pohl (2006), analyzing the German socioeconomic panel study (2000–2003), found no statistically significant difference in the chances for immigrant students to achieve a higher educational degree in relation to the age of migrating to Germany.

In Chap. 4B, I will investigate if I can find relationships in TIMSS between the age of immigration and the immigrants' achievement.

Differences Between Girls and Boys

Another basic factor of differentiation regarding student achievement that is examined in research is the sex of the students. The literature about the achievement differences of boys and girls – especially in mathematics and in science – is numerous.

Even at the beginning of the TIMSS cycles, right after the first results from the first cycle were published, the TIMSS 1995 study center at Boston College published a separate report about the differences between girls and boys (Mullis, Martin, Fierros, Goldberg, & Stemler, 2000). They found: "The gender differences in achievement in both curriculum areas widened at the upper grades. Thus, while males in the fourth grade had higher achievement than females in only some countries, by the final year of secondary school gender differences in performance were pervasive – with males having significantly higher achievement than females in both curriculum areas in almost every TIMSS country" (p. 30). This is in line with most other research.

In TIMSS 2007, however, the picture changed quite substantially. Only in seven participating countries (Lebanon, Australia, Syria, El Salvador, Tunisia, Ghana, and Columbia) and two benchmark participants (British Columbia and Ontario) boys scored significantly better than girls. But in 16 countries (Lithuania, Malaysia, Egypt, Bulgaria, Singapore, Botswana, Romania, Cyprus, Jordan, Kuwait, Saudi Arabia, Thailand, Bahrain, Palestine, Qatar, and Oman), girls scored statistically significant higher than boys in mathematics which resulted in even the international average for girls being statistically significant higher than that for boys (Mullis et al., 2008, p. 59).

For immigrant students, Dronkers and Kornder (2013, p. 1) found: "The principal conclusion was that female migrant pupils have higher reading and math scores than comparable male migrant pupils and these gender differences among migrant pupils in reading and math scores are larger than among comparable native pupils." The major reason for this was that "The majority or at least a large minority of migrants to OECD countries move from societies with less gender equality to societies with a more equal power balance between the sexes" (p. 3). Although they researched the achievement differences for immigrant students, interestingly they did not consider the different percentages of immigrant boys and girls as reported in Chap. 4B of this publication.

Language Difficulties

A precondition for students to follow lessons and learn effectively is the mastery of the language of instruction. The IEA has stated: "Reading literacy is one of the most important abilities students acquire as they progress through their early school years. It is the foundation for learning across all subjects, it can be used for recreation and for personal growth, and it equips young children with the ability to participate fully in their communities and the larger society" (Mullis, Kennedy, Martin, & Sainsbury, 2006, p. 1).

Language is not only needed for learning, but also almost all assessments of what is learned in school require the mastery of the language of the test. Mullis et al. used the opportunity that in 2011 PIRLS as well as TIMSS took place and that several countries participated in both surveys. Some countries chose the option of assessing the same students in TIMSS and PIRLS in grade four which lead to a database that includes mathematics, science, and reading scores for the students in these countries. Mullis et al. hypothesized: "Students with high reading ability would not be impacted by the level of reading demand in the items. That is, the best readers would score similarly on TIMSS items regardless of the degree of reading demands" (Mullis, Martin, & Foy, 2013, p. 2). Analyzing the data, they discovered that "The average mathematics achievement of the best readers did not vary much by level of reading demand whereas the average mathematics achievement of the least proficient readers was higher on the items with low reading demands than on the items with medium and high reading demands. While the poorest readers consistently achieved at a lower level in mathematics than the best readers, they were additionally disadvantaged on the mathematics items that required more reading" (Mullis et al., 2013, p. 15). The results for science were similar, leaving me to conclude that especially students with low reading abilities had difficulties answering mathematics and science items correctly and that their overall score is impacted by their language difficulties.

Language difficulties are also often considered as the main factor for the lower performance of immigrant students (see, e.g., (Buchmann & Parrado, 2006)). The OECD (2013b, p. 80) stated that "Most disadvantaged migrants are those not speaking the host-country language" and "… language support should be a priority in migrant education policy" (2010a, p. 46). Furthermore the OECD notes: "In most countries, there is a relative under-representation of students that speak the language at home among low achievers,…" (2011a, p. 31).

Socioeconomic Background and Parental Education

Research showed that a major portion of achievement differences can be attributed to the students' socioeconomic background (see, e.g., (Sirin, 2005)). The socioeconomic status (SES) is usually defined as the "relative position of an individual or family within a hierarchical social structure, based on their access to, or control over wealth, prestige, and power" (see, e.g., (Mueller & Parcel, 1981)). The SES is usually separated into the four domains: economical capital, social capital, symbolic capital, and cultural capital (Bourdieu, 1983). The SES is measured differently in different international large-scale assessment studies. TIMSS and PISA only have student questionnaires, whereas PIRLS also has a parent questionnaire that includes questions related to SES. A review on the different measures is presented by Brese and Mirazchiyski (2010).

After the publication of the so-called Coleman et al. (1966) report, it has been a well-known fact in educational research that the socioeconomic background of the parents correlates strongly with the achievement outcomes of their children. Consequently, most educational studies try to measure the SES of students and include it in statistical models as a control variable. Sirin (2005, p. 434) found: "Of 64 independent student-level studies, 62 reported information about the source of SES data."

In educational research, the SES is usually measured by different indicators or resources. Hattie (2008, p. 61) wrote: "Such resources refer to the parental income, parental education, and parental occupation as three main indicators of SES." Later Hattie refers to the effect sizes found by Sirin (2005, p. 434): "A weighted ANOVA revealed that the average SES was .28 for parental occupation, .29 for parental income, and .30 for parental education." We see that the parental education has the biggest effect size.

For the TIMSS 2007 data, Mullis et al. (2008, p. 145) explained: "Nonetheless, Exhibit 4.1 makes it clear that higher levels of parents' education are associated with higher average mathematics achievement in almost all countries." I expect similar findings for my analysis of the education of immigrant student's parents and its relation to the student's achievement.

Hansson and Gustafsson (2010) investigated the invariance of socioeconomic status measures of immigrant and nonimmigrant student groups in Sweden. They found significant differences especially in the number of books at home for the different student groups and suggest to always consider the migration status of the students when analyzing this variable. Postlethwaite noted the same when analyzing data from the IEA Reading Literacy Study: "The only variable that provided a surrogate measure of socioeconomic level and was positively correlated with reading achievement in all countries was *number of books in the home*" (Postlethwaite & Ross, 1992, p. 22).

For immigrants, the OECD (2010a, p. 37) observed: "Although immigrants are a very heterogeneous group, significant proportions of immigrant students come from less advantaged socio-economic backgrounds" and also "In general, students with an immigrant background are socio-economically disadvantaged, and this explains part of the performance disadvantage among these students" (2010c, p. 71).

Attitudes and Aspiration

In the next section, I focus on the students' attitudes and aspirations. Students' attitudes and aspirations can be defined as outcome of the educational process similar to achievement. A positive outcome of education is that the students feel comfortable in school, have positive attitudes about the subjects taught in school, and have high educational aspirations. But at the same time, students' attitudes and aspirations can also be seen as process variables that influence the achievement outcome positively. Mullis et al. (2007, p. 139) stated: "Positive student attitudes toward reading and a healthy reading self-concept are major objectives of the reading curriculum in most countries. Students who enjoy reading and who perceive themselves to be good readers usually read more frequently and more widely, which in turn broadens their reading experience and improves their comprehension skills."

Alvernini et al. (2010) used PIRLS 2006 data to show that student's attitude towards reading has a moderator effect on the relationship between school, teacher, and home background and on student's reading achievement in Italy.

Osborne (2003, p. 1054) separates the attitude towards science but also mathematics into different components such as anxiety towards science, the value of science, self-esteem at science, motivation towards science, and enjoyment of science. He also suggests distinguishing between science in general and school science since students' attitudes towards science in general have a tendency to be more positive than towards school science. Based on his literature review he concludes: "Within all of the literature, there is some disagreement about the nature of the causal link and whether it is attitude or achievement that is the dependent variable. The essential premise permeating much of the research is that attitude precedes behavior" (p. 1072). Furthermore, he emphasizes the importance of students' positive attitudes towards science and mathematics to change the trend of reduced interest in science-related further education and careers. "Its current importance is emphasized by the now mounting evidence of a decline in the interest of young people in pursuing scientific careers." (p. 1049)

Mullis et al. (2008, p. 173) stated that "Developing positive attitudes toward mathematics is an important goal of the mathematics curriculum in many countries." But they also understand that the attitudes are not only a goal in itself but might lead to higher achievement. "In addition to having a positive attitude toward mathematics, students may be more attracted to mathematics and more motivated to learn it if they perceive mathematics achievement as advantageous to their future education and the world of work." (p. 174) They found that "Average mathematics achievement was highest among students at the high index level [of students' positive attitudes towards mathematics]" (p. 174).

Kiamanesh and Mahdavi-Hezaveh (2008) present a comprehensive literature review on attitudes of students towards mathematics.

Mata et al. (2012) researched the interconnectedness of students' attitudes towards mathematics and mathematics achievement. The results of their longitudinal study in Portugal show that the attitudes of students towards mathematics declined throughout the school career.

Of special interest are the differing attitudes towards mathematics by boys and girls. Meece et al. (2006, p. 367) explain that "In general, boys tend to have positive achievement-related beliefs in the areas of mathematics, science, and sports while girls report show more favorable motivation patterns in language arts and reading." Their review includes the different motivational theories about acquiring self-expectancies and attitudes towards mathematics and science between girls and boys. Drawing conclusions from the US NAEP assessment, they concluded that: "Specifically, girls reported less interest in pursuing mathematics and science careers lower participation in math- and science-related extracurricular activities, and less confidence in their mathematics abilities than did their male counterparts" (p. 366).

Mata et al. (2012, p. 9) noted in their longitudinal study in Portugal "… a systematic decline in attitudes towards mathematics along schooling" and especially for girls a progressive decline in attitudes which they explain by gender stereotypes.

Differences Between Country Regions

Schools can be located in urban areas or in more rural areas. The situation of the students and of the schools in the different community sizes can differ significantly. In some countries, rural schools are much smaller and offer fewer courses than urban schools; in others, it is more difficult to attract good teachers in rural schools. This can also have consequences for the quality of education and educational outcome.

For the educational outcome, Beaton et al. came up with interesting results for the Russian Federation in their analysis of TIMSS 1995 data: "There are great differences in mean student achievement in relation to school location and student gender. It is important to note that these differences do not occur in relation to mathematics education. The school location factor, in particular, has a significant impact on science achievement in the Russian Federation. The farther a school was from a center of a region, the lower the mean science achievement on TIMSS. The lowest achieving schools were located in rural areas" (Robitaille & Beaton, 2002, p. 181).

For the educational outcome in Latvia, Johansone (2010, p. 1) detected: "A significant part of the achievement variance can be explained by performance differences between urbaan and rural school communities," explaining that "Poor equity of achievement in Latvia's primary education is a problem of segregation by socioeconomic status, and the urbanization effect is significant mostly because the segregation is more obvious in the rural areas of the country" (p. 16).

In an international perspective, the educational outcome was addressed in a report about effective schools based on TIMSS 1995 data: "Although in several countries greater percentages of students in low achieving schools were located in urban areas, which supports the idea that urban schools are often disadvantaged, only for Scotland in science was the difference statistically significant. In contrast, in seven countries, Austria, Cyprus, Hungary, Iran, Korea (mathematics only), and the Russian Federation, significantly greater percentages of students in the high-achieving schools were in schools located in urban areas. Of these countries, both Iran and the Russian Federation have large tracts of remote areas, and the difference between urban and rural can be very marked" (Martin, Mullis, Gregory, Hoyle, & Shen, 2000, p. 46).

Even earlier, Postlethwaite and Ross (1992) examined effective schools in reading with the IEA 1991 Reading Literacy Study data. They found: "The more effective school has a community context that tends to be urban and which features ready access to books through the availability of a public library and a local bookstore. In addition, further education opportunities are offered beyond primary school because of the proximity of a secondary school and a higher education institution" (p. 42).

Further, the OECD (2010b) addressed the issue of rural schools in several projects. They found the education in rural areas problematic: "Education is the cornerstone of rural development but, delivering education to sparsely populated areas presents with some challenges. Some institutions suffer from problems of limited capacity, poor quality, relevance and limited public funding. There is often a mismatch between the education offered and the needs of the rural regions" (p. 91).

Especially the recruitment was found to be problematic in several countries: "In Australia, schools in remote and rural areas have been experiencing difficulties in attracting and retaining teachers. To encourage teachers to teach and remain in those areas beyond the minimum required service period, special incentives and teacher education programmes are offered in most States, as illustrated by Queensland and New South Wales" (Organisation for Economic Co-operation and Development, 2005, p. 51). Some countries pay a special allowance to teachers in rural schools or even make it an obligation for teachers after the initial teacher training to teach in rural communities, e.g., in Korea.

But not only the education of students in rural areas proves to be problematic in some countries; the in-service training of teachers is also an issue. This is due to the fact that the offerings are limited in some cases, or that the funds allocated to rural schools were less generous in some countries. The OECD (2003a, p. 70) reported on the topic: "Participation in continuing education and training is considerably more extensive among teachers in urban municipalities when compared with remote rural municipalities." The lack of teachers' participation in in-service training can also have an impact on the teaching quality and consequently on students' outcome.

I conclude that the education in rural areas might face challenges in certain countries. In some countries it is difficult to fill teaching positions and keep good teachers in rural schools. In general the infrastructure in rural areas is less developed and thus makes access to libraries but also to further education difficult for students.

School Attendance

Attending school regularly is important for the learning success of all students. As Büchel et al. (2001, p. 151) wrote: "Attending school is important for two reasons. First and most obvious, school helps children to acquire learning skills and information on a wide range of subjects. Second, and in many ways just as important, formal schooling provides the forum through which children develop social skills, learning to be independent and to relate to non-family members in a group-based setting".

In most countries, attending school is obligatory for children of a certain age, although it is not given in all developing countries. This is especially true for countries involved in violence. UNESCO (2013a, p. 1) reported: "Globally, the number of children out of school has fallen, from 60 million in 2008 to 57 million in 2011. But the benefits of this progress have not reached children in conflict-affected countries. These children make up 22 % of the world's primary school aged population, yet they comprise 50 % of children who are denied an education, a proportion that has increased from 42 % in 2008 [...]. Of the 28.5 million primary school age children out of school in conflict-affected countries, 12.6 million live in sub-Saharan Africa, 5.3 million live in South and West Asia, and 4 million live in the Arab States. The vast majority – 95 % – live in low and lower middle income countries."

But also in developed countries, not all students do always attend school. Büchel et al. (2001) analyzed longitudinal data from West Germany and found that regular school attendance is strongly related to the parental income. Especially children with lower socioeconomic background have a higher tendency for lower school attendance.

The OECD (2003b, p. 8) has dedicated a separate report to school attendance after PISA 2000 which considers the relationship between school attendance and achievement somewhat proven: "There is a more distinct, but still weak association between participation and performance among individuals. However, in both cases there is a moderately strong association between schools in which students are engaged and those with good overall student results." Interestingly, the association between school attendance and student achievement was more prominent at school level than at the individual student level.

In the same report, the OECD (2003b) presented some student background characteristics that are associated with the students' participation in school: "The quarter of students with least favourable backgrounds, measured by parental occupation and education, are: … 26 % more likely than students of medium social background to have low participation, on average in OECD countries…" (p. 11). Also single-parent families seem to have a more pronounced problem with school attendance. The OECD also noted: "Students with single parents are: 40 % more likely than other students to have low participation, on average in OECD countries…" (p. 13).

TIMSS puts special emphasis on measuring the students' school attendance. A set of three questions is directed at school principals concerning the seriousness of students' absenteeism, arriving late at school, and skipping classes. An indicator variable is created that distinguishes between schools with high school attendance, medium school attendance, and low school attendance (Foy & Olson, 2009). The low category is defined as having at least two out of the three issues as serious problems. In several countries, there is a statistically significant higher percentage of first-generation immigrants going to schools with a low school attendance than native students. Mullis et al. (2008, p. 326) found that school attendance can be a problematic issue in schools as achievement results usually relate positively to school attendance: "Attendance problems appear to be more serious at the eighth grade than at the fourth, with an average of 21 percent of the students at the high index level compared with 43 percent at fourth grade, and 20 percent at the low level compared with just 7 percent at fourth grade." For the relation to achievement they found: "Average mathematics achievement was highest among students at the high index level (478), next among those at medium level (471 points), and lowest among those at the low level (432 points)" (p. 326).

The Effect of School Resources

There is a general tendency in recent education policy research to focus on teacher factors when analyzing how to improve education. For example, the OECD (2005, p. 23) stated, "The research indicates that raising teacher quality is perhaps the policy direction most likely to lead to substantial gains in school performance." School factors also play an important role in education, e.g., a shortage in teaching materials can impact the teaching negatively. Already in the beginning of the 1990s when Postlethwaite and Ross (1992, p. 30) did research on effective schools, they concluded that, "In all cases, the more effective schools had more resources than less effective schools."

The OECD (2011a) analyzed the situation of resilient students and tried to find situations that impact the achievement of disadvantaged students positively. They found: "...resilient students enjoy better resources than disadvantaged low achievers" (p. 59). This is another hint that better resourced schools can impact student achievement – especially in the case of immigrant students who are, according to the OECD, as being overrepresented in the group of disadvantaged students.

But there is a caveat regarding the impact of school resources on student achievement and the results based on cross-sectional data. Postlethwaite indicated there is a clear link between students' background and school resources: Students from privileged areas tend to attend well-resourced schools.

PISA 2009 revealed that "...differences in the socio-economic background of schools in many countries make it difficult to provide equity in learning opportunities for students with an immigrant background, inequality in the distribution of resources does not seem to mediate the performance gaps between students with and without an immigrant background except in a small number of countries" (OECD, 2010c, p. 81).

The School Climate

Comprehensive research has been conducted about the school climate and how it relates to student achievement. Although there is no definition of the term "school climate," there is a common understanding of its dimensions and an agreement that a positive school climate has a positive impact on students' learning. It can be a good measure of risk prevention to influence the school climate positively as shown by Freiberg (1998, p. 1): "The elements that make up school climate are complex, ranging from the quality of interactions in the teachers' lounge to the noise levels in hallways and cafeterias, from the physical structure of the building to the physical comfort levels (involving such factors as heating, cooling, and lighting) of the individuals and how safe they feel. Even the size of the school and the opportunities for students and teachers to interact in small groups both formally and informally add to or detract from the health of the learning environment. The support staff—cafeteria workers, bus drivers, custodians, and office staff—add to the multiple dimensions of climate. No single factor determines a school's climate. However, the interaction of various school and classroom factors can create a fabric of support that enables all members of the school community to teach and learn at optimum levels. Further, making even small changes in schools and classrooms can lead to significant improvements in climate."

Cohen (2009, p. 1) uses the following definition of school climate: "School climate refers to the quality and character of school life. School climate is based on patterns of students', parents' and school personnel's experience of school life and reflects norms, goals, values, interpersonal relationships, teaching and learning practices, and organizational structures. A sustainable, positive school climate fosters youth development and learning necessary for a productive, contributing and satisfying life in a democratic society. This climate includes: norms, values and expectations that support people feeling socially, emotionally and physically safe.

People are engaged and respected. Students, families and educators work together to develop, live and contribute to a shared school vision. Educators model and nurture attitudes that emphasize the benefits and satisfaction gained from learning. Each person contributes to the operations of the school and the care of the physical environment."

Shumow and Lomax (2001, p. 106) concluded in their studies: "The findings demonstrated that positive climate was linked to greater parent involvement and to higher educational aspirations among students, which were, in turn, linked to reports of safer schools. This indicates that improvement of a school's climate might be an effective way to promote parent involvement and to increase school safety." This shows the importance of a positive school climate and its effects on other aspects in the students' learning environment that have the potential to positively influence the academic success as well as further dispositions.

A summary of school climate research is presented in Center for Social and Emotional Education (2009).

In TIMSS, the school climate is assessed in the teacher and in the school principal questionnaires. The principals as well as the teachers were asked to rate from "very high" to "very low":

- Teacher's job satisfaction
- Teacher's understanding of the school's curricular goals
- Teacher's degree of success in implementing the schools' curriculum
- Teacher's expectations for student achievement
- Parental support for student achievement
- Parental involvement in school activities
- Student's regard for school property
- Student's desire to do well in school

Then an index with three levels was calculated. The highest level was assigned if answers averaged to high or very high, low was assigned if the average was low or very low, and the middle category was generated for answers in between (Mullis et al., 2008, p. 355).

At the teacher level, Mullis et al. (2008) found: "Average mathematics achievement was positively related to teacher's perception of school climate at both fourth and eighth grades, with average achievement higher among students at the high index level and lower among students at the low level of the index" (p. 357) and similar for the principal ratings: "At both fourth and eighth grades, average mathematics achievement was highest among students at the high level of principals' perception of school climate index (487 points and 473 points, respectively), next highest at the medium level (471 and 450 points, respectively), and lowest at the low level (441 and 428 points, respectively)" (p. 356).

Kozina et al. (2010) analyzed TIMSS advanced data and found a positive correlation between the school climate and the student achievement in the schools. The highest correlation was found for the school climate as reported by the school principal, the next highest for the school climate reported by the teacher and the lowest for school climate as reported by the students.

School Safety

Feeling safe in school is a related aspect that has shown to be important for students' academic success.

Bowen and Bowen's (1999, p. 337) survey of more than 2000 students in the United States revealed that "The regression results suggest a relationship between environmental danger and school performance that is supported by theory and other research."

Chen and colleagues (1997) did a longitudinal study with more than 500 students in Shanghai and found a clear connection between the students' experience of safety in the school and their academic success. Interestingly the causality is not unidirectional, but safety and academic success seem to influence each other: "In summary, it was found that academic achievement predicted children's social competence and peer acceptance. In turn, children's social functioning and adjustment, including social competence, aggression-disruption, leadership, and peer acceptance, uniquely contributed to academic achievement" (p. 524).

Perŝe et al. (2008, p. 6) analyzed the TIMSS 2003 data for Slovenia, concluding: "National analyses for Slovenia show important associations of educational achievement and negative school factors. The results show significant differences in math and science achievement between the following pairs of groups: students whose things were stolen in the last month and students whose things were not stolen; those who were physically harmed and those who were not; those who were forced into activities they did not choose and those who were not; those who were called names and those who were not; those who were left out of activities and those who were not. In all of these groups students who experienced aggressive behavior scored lower in math and science, both in the 4th and 8th grade."

TIMSS 2007 includes an index on safe and orderly schools that is covered at teacher and student levels. This index includes teacher level variables such as "this school is located in a safe neighborhood," "I feel safe at this school," and "this school's security policies and practices are sufficient" for the teacher level index. Probably, more relevant to the students is what is covered in the student-level safety and orderly index. This index includes the aspects:

- Something of mine was stolen.
- I was hit or hurt by other student(s) (e.g., shoving, hitting, kicking).
- I was made to do things I didn't want to do by other students.
- I was made fun of or called names.
- I was left out of activities by other students.

The index has three values: high, medium, and low. Students were assigned the high value if all five statements were answered negatively. They scored "low" if at least three statements were answered "yes" and "medium" in all other cases (Mullis et al., 2008, p. 363).

Class Size

One of the most discussed factors that can influence student achievement is the class size or, closely related, the student-teacher ratio because it is easy to measure and can easily be influenced by policy makers. The question whether class size has an impact on student achievement is discussed in many research and policy papers and the conclusions differ vastly. Economists tend to argue that class size has no or little effect (Wößmann & West, 2006) and that public expenditure can be allocated in much better ways. "Simple cost-benefit considerations suggest that even in Iceland, where class-size effects are statistically significant, the future income gains induced by increases in educational performance are unlikely to offset the costs induced by reductions in class size." (Wößmann, 2007, p. 17) The usually large classes common in Asian countries such as Korea, Japan, or Chinese Taipei together with their high achievement in international large-scale assessment studies are used as arguments that class sizes do not matter. On the other hand, case studies seem to show that class size does matter (Haimson, 2000). Probably the most influential study in this area is the student-teacher achievement ratio project (STAR) that was conducted in Tennessee in the 1980s. The STAR project was an experimental study that revealed: "There is a consistent and fairly large scaled score difference favoring the small class over the regular class at each grade" (Word et al., 1990, p. 26).

In TIMSS, mathematics (and also science) teachers were asked about the size of the class in which they are teaching the sampled students. Then an indicator was computed. For the grade eight students, the indicator distinguishes between small classes with one to 19 students, medium-sized classes with 20–32 students, and large classes with 33 or more students. Regarding the relationship to achievement, the TIMSS 2007 mathematics report stated: "Because countries have a variety of policies, practices, and realities determining class sizes, the relationship between class size and achievement is extremely difficulty to disentangle" (Mullis et al., 2008, p. 272) and "The complexity of this issue is evidenced in the TIMSS 2007 results showing a curvilinear relationship, on average, between class size and mathematics achievement at both the eighth and fourth grades" (p. 273).

Homework

The effect of homework on student achievement is widely discussed. Hattie (2008) counted 161 studies on this topic. He concluded in his meta-analysis that homework has only little impact on student achievement and that the impact depends heavily on the kind of homework. He states: "It is clear that, yet again, it is the differences in the teachers that make the difference in student learning. Homework in which there is no active involvement by the teacher does not contribute to students learning, …" (p. 236).

Similarly Cooper et al. (2006) came to the conclusion that homework has only little effect on student achievement and that the lower the grades, the lesser the effect of homework. On the other hand, Trautwein et al. (2002) analyzed a data set of about 200 students regarding homework assignments and its relation to the mathematics achievement of the students. They found that "… our data support the assumption that homework is substantially related to achievement gains in mathematics. In our study, the explained variance after controlling for several entry and system variables was about 8% at the class level" (p. 43).

Interestingly Hattie (2008, p. 235) also concluded from his meta-analysis that "The effects [of homework] are greater for higher than for lower ability students…." In contrast, Trautwein et al. (2002, p. 45) found: "This interaction effect indicates that low-achieving students gain more than high-achieving students from extensive homework assignments." In Chap. 4A we will see that students with an immigrant background are mostly lower-achieving students. Consequently, one would expect that, following Hattie, the effect of homework would be smaller for students with an immigrant background than for native students, or, following Trautwein et al., that the effect of homework would be more noticeable for students with an immigrant background than for native students.

In TIMSS, students were asked about the frequency of homework in mathematics and in the science subjects, and also how much time they spend on doing the homework. Relating the students' answers on "time on homework spent" to achievement, however, entails the difficulty that even if the assignment of homework had a positive impact on student achievement, this is hidden by the effect that lower-performing students need more time finishing the tasks. Ronnig (2010) analyzed the Norwegian TIMSS 2007 data with respect to homework and especially looked for differences concerning students with a different socioeconomic background. He concluded: "At the same time, it is also found that if pupils from lower socio-economic backgrounds spend time on homework, they actually spend more time on it than pupils from higher socio-economic backgrounds" (p. 23). He hypothesized that "they may need more time in to complete their homework if they find the homework more difficult than pupils from higher socio-economic backgrounds. Also more time spend on homework can reflect problems related to motivation, frustration and concentration (Trautwein and Köller, 2003). On the other hand, more time spend on homework may also reflect high educational ambitions, regardless of socioeconomic background" (Ronnig, 2010, p. 23).

All these potential effects make the analysis of the student-level data difficult. In TIMSS mathematics teachers are asked how much homework they assign to their students, and the answers were transformed into an indicator of "teachers' emphasis on mathematics homework." "Students in the high category had teachers who reported giving relatively long homework assignments (more than 30 min) on a relative frequent basis (in about half of the lessons or more). Students in the low category had teachers who gave short assignments (less than 30 min) relatively infrequently (in about half the lessons or less). The medium level includes all other possible combinations of responses" (Mullis et al., 2008, p. 302). The results reported in the international TIMSS 2007 mathematics report revealed that "There was little relationship between teacher's emphasis on homework and mathematics achievement" (Mullis et al., 2008, p. 302). As can be seen in Chap. 4D, students.

Homework does not only relate to student achievement but can also relate to students' attitudes. Cooper et al. (2006) investigated effects of homework on non-academic outcome. They concluded, "Five studies that presented correlations between the amount of time students spent doing homework and student attitudes revealed a significant positive relationship using a fixed-error model" (p. 52). This means that students who spend more time on mathematics homework have a more positive attitude towards mathematics.

Peer Effects

There are different peer effects researched and described in the literature. In terms of achievement, there are different effects that can be observed when students are grouped together with high- or low-ability students, respectively. On the one hand, there is the "big fish in little pond" effect introduced by Marsh and Parker (1984) which implies that students who are performing better than their peers have a higher self-esteem regarding their own abilities. On the other hand, Marsh and Parker also describe the assimilation or "reflected glory effect" that implies the positive effect on self-esteem of students who feel that they are selected together with high-performing students.

These two effects occur independent of each other and with different effect sizes dependent of the cultural context as researched by Marsh et al. (2000). Higher self-esteem then has a positive effect on the achievement, as also shown by Byrne (1996) or Pajares et al. (2000). Jen and Chien (2008), however, argue that a relatively high self-concept has a positive effect on the achievement in the same subject but a nega-tive effect on the achievement in other subjects. The academic aspirations of the students as well as the average academic aspirations of the class also have a positive effect on the students' achievement as shown by Martin et al. (2000).

Of particular interest is a public discussion on the effect of nonnative peers in several countries. Especially parents are concerned about the number of students with an immigrant background in their children's classes. In Germany the German Foundations' Council of Experts for Integration and Migration ("Vorwurf vom Integrationsrat," 2012) elaborated that "Viele Eltern, die ihren Sprösslingen einen erfolgreichen Start ins Schulleben ermöglichen wollen, schicken sie auf eine Grundschule mit möglichst wenigen ausländischen Kindern – und schaden damit dem deutschen Bildungssystem" ("Many parents who want to provide to their off-springs a successful start in school life send them to a primary school with prefera-bly few foreign children – and thus harm the German education system"). The experts found that parents are choosing the school for their children also based on the percentage of immigrant students in the school, which leads to a higher segrega-tion between native and immigrant students in schools than in the neighboring communities. This is a pressing topic from the public's perspective and consequently intense research is ongoing about peer effects for immigrant students.

A meta-analysis on this topic is carried out by Dalit (2011, p. 28) who explains: "The considerable growth of the share of immigrant students which has occurred over the last decade has contributed to raise the concern within large sectors of the public opinion that immigrant children would have a negative influence on the school performance of natives. However, this concern does not seem to be empirically well-founded. The analyses carried out in this paper point to the existence of negative effects of the concentration of immigrant students on peer performance; yet, these effects are small and heterogeneous. As regards Italian language test, the concentration of first generation immigrant students appears to influence immigrants more than natives. Among natives, while low socio-economic background children may somewhat suffer from a large share of immigrant background classmates, children of higher background do not; on the contrary, in some cases they even seem to benefit from the presence of immigrants."

Moreover, Conger (2010) found diverse results when analyzing different cohorts of public high school students from Florida. He detected that immigrant students caused general negative peer effects for tenth graders, but there were also spillover effects for a subgroup of eighth graders when they were schooled with immigrant students early in their school career: "It turns out that the immigrant peers in the schools attended by the sub-population of students are a high performing group – foreign-born students who arrived to the U.S. when they were younger" (p. 20).

On the other hand, Yeung (2011) determined in his publication on peer effects of immigrant students from East Asia and the Dominican Republic in New York very clear peer effects disadvantaging students with immigrant peers. He concluded: "The results of both the East Asian and Dominican regressions are conspicuous in their similarity. Both East Asian and Dominican immigrant composition have negative and significant effects on achievement. These findings are consistent with most of the research on immigrant composition effects (e.g. Cho, 2011, DiPaolo, 2010, Friesen and Krauth, in press, Gould, et al., 2005). Bother types of immigrant composition have stronger effects in mathematics. East Asian immigrant composition has negative effects on East Asian immigrant children as well as other children. Likewise, Dominican immigrant composition has negative effects on Dominican immigrant children as well as other children. The results collectively suggest there is a negative effect of immigrant composition that is independent of ethnicity and culture" (p. 158).

Ohinata and van Ours (2011) examined IEA TIMSS and PIRLS data for the Netherlands and could not find a negative spillover effect of immigrant students on their native peers. But: "Immigrant children themselves experience negative language spill-over effects from a high share of immigrant students in the classroom but no spill-over effects on maths and science skills" (p. 21). They suggest relocation of immigrant students to other schools because they "… might benefit from such reallocation as their language skills might improve once they can interact more intensely with the native Dutch children" (p. 21). Interestingly Netten (2010), on the other hand, found negative peer effects when analyzing IEA PIRLS data from the Netherlands.

Andersen and Tomsen (2011) analyzed a data set with 40,000 Danish grade nine students. They even suggested to policy makers to limit the percentage of immigrant students in schools to 50 % and to relocate students to other schools when this threshold is reached to avoid negative peer effects on native Danish students.

In summary results from studies of peer effects caused by immigrant students are quite diverse, but there seems to be some evidence that a high percentage of immigrant students in a class influences their own achievement – especially in language – negatively. More ambiguity appears to exist with regard to the effects detectable in other subjects as well as the effect on native students.

Chapter 3
Data and Methods

Abstract This chapter describes the TIMSS assessment which was designed to measure mathematics and science abilities at the end of primary education and at the beginning of secondary education. As a curriculum-based assessment, TIMSS assesses students in content and cognitive domains in the form of multiple-choice as well as open-ended items. The chapter describes the study's background questionnaires and the grade-based, matrix-sampling design. It explains the TIMSS scoring where Item Response Theory (IRT) is used to analyze the test data and to assign achievement scores to students. The format of the data and the procedures used make it necessary to use sophisticated methods in the analysis. The chapter also presents the IEA International Database Analyzer (IDB), a tool which helps researchers calculate statistics and standard errors of differences. The deliberations also explain that the focus is on the mathematics scores because the influence of language on achievement is low, and mathematics classes are in most countries used in the sampling. The chapter also helps to give meaning to differences in score points.

Keywords TIMSS • Item response theory • IRT • Mathematics achievement

After presenting the research questions that guide this publication, reviewing the literature, and describing the current stage of the, in this chapter I will specify the data and methods used in the analysis to follow.

The quantitative part of this research will make use of IEA TIMSS data. TIMSS assesses mathematics and science achievement of grade four and grade eight students every 4 years starting in 1995 in up to 66 countries. The study gathers an extensive amount of background information from the students, their teachers, and school principals.

TIMSS

TIMSS is a research project launched by the International Association for the Evaluation of Education Achievement (IEA). IEA is an international nonprofit research organization that was founded in 1958. The members of IEA are educational research institutes, ministries of education, and universities. IEA was founded

© Springer International Publishing Switzerland 2016
D. Hastedt, *Mathematics Achievement of Immigrant Students*,
DOI 10.1007/978-3-319-29311-0_3

with the idea of conducting large-scale assessments internationally in order to learn from differences and similarities between countries. As A. W. Foshay, one of the founding fathers of the IEA, stated: "If custom and law define what is educationally allowable within a nation, the educational systems beyond one's national boundaries suggest what is educationally possible" (IEA, 2011). Until today, IEA has conducted more than 30 international educational studies.

In 1998 IEA developed the idea of conducting an international study on mathematics and science abilities of students. The study was based on previous IEA studies – namely, the First International Mathematics Study (FIMS) conducted in 1964, the Second International Mathematics Study (SIMS) conducted between 1980 and 1982, the First International Science Study (FISS) conducted in 1970 and 1971, and the Second International Science Study (SISS) conducted in 1983 and 1984. All studies were developed by a designated international study coordination center together with researchers from the participating countries and international experts. For the processor studies, the international study centers were established at the University of Stockholm, Sweden, and the University of Hamburg, Germany, respectively. The international study center for the Third International Mathematics and Science Study (TIMSS) was located at the University of Vancouver, Canada, until 1995 and from then on at the Lynch School of Education at Boston College, USA.

After the initial TIMSS in 1995, a TIMSS-Repeat study (TIMSS-R) was launched in 1995. TIMSS-R's focus was narrower: whereas in TIMSS 1995 two adjacent grades were assessed per population (see TIMSS sample below), TIMSS-R assessed only the upper grade of population 2 – the eighth graders. The number of participating countries[1] decreased from 45 to 38. Many countries and researchers expressed an interest in continuing the TIMSS data collection every 4 years. As a consequence, the Third International Mathematics and Science Study turned into the Trends in International Mathematics and Science Study and is since then conducted every 4 years assessing grade four and grade eight students in a growing number of countries.

The TIMSS Assessment

TIMSS 1995 was designed to measure mathematics and science abilities at the end of primary education through a sample of grade three and grade four students and at the beginning of secondary education through a sample of grade seven and grade eight students. TIMSS defined these populations as "The two adjacent grades with

---

[1] It must be mentioned here that IEA defined a country in terms of an educational system. For example, the French and the Flemish parts of Belgium have different curricula and educational policies. Consequently from the beginning of IEA, both entities became separate IEA members. Results from the Flemish and the French part of Belgium are always reported separately in all IEA reports. The same is true for England, Ireland, and Scotland.

the largest proportion of 9-year-olds at the time of testing (third and fourth grades in many countries)" and "The two adjacent grades with the largest proportion of 13-year-olds at the time of testing (seventh and eighth grades in many countries)" (Martin & Kelly, 1997a, p. 1).

The grade three and grade four cohorts were named lower and upper grade of population 1; the seven and eight grade cohorts were named lower and upper grade of population 2. The concept of TIMSS is a three-strand model that takes the perspective of the intended curriculum, the implemented curriculum, and the attained curriculum (Martin & Kelly, 1997a, Chapter 1.2). This concept is taken from the IEA SIMS and investigates "what society would like to see taught," "what is actually taught in the classroom," and "what the students learn" (Martin & Kelly, 1997a, p. 3). Extensive analyses of all the participating countries' curricula and text books were performed.

The TIMSS assessment is a curriculum-based assessment[2] which means that the assessment is evaluated by experts and matched to the national curricula of participating countries. Based on this idea of assessing what is taught in the various countries in school by the seventh and eighth grade, the TIMSS 1995 assessment includes for population 2 items in six mathematics domains: "fraction and number sense; measurement; proportionality; data representation, analysis and probability; geometry and algebra" (Beaton, Mullis, et al., 1996, p. 1). The population 2 science assessment included items from five content dimensions: "earth science, life science, chemistry and environmental issues and the nature of science" (p. 1).

Items were developed by subject matter experts and country representatives of participating countries. The items were piloted, used in a field trial, and thoroughly reviewed before entering in the final assessment (Martin & Kelly, 1997b, Chapter 2). To cover all domains adequately, for mathematics 125 multiple-choice items, 19 short-answer items, and seven extended response items were included in the final assessment and for science 102 multiple-choice items, 22 short-answer items, and 11 extended response items (Martin & Kelly, 1997b, tables 3.13 and 3.14). "The design thus provides 396 unique testing minutes, 198 for science and 198 for mathematics" (p. 16). Since not all items could be given to any single student, the items were grouped together into item clusters, and these clusters were assembled to eight different but overlapping test booklets. Each of the booklets required 90 min of testing time.

For all cycles, TIMSS is a curriculum-based assessment (see "The TIMSS curriculum model" (Mullis, Martin, Ruddock, et al., 2005, p. 4). The assessment design includes the content domains number, geometric shapes and measures, and data display for grade four; number, algebra, geometry, and data; and chance for grade eight. The cognitive domains are knowing, applying, and reasoning (see (Mullis, Martin, Ruddock, O'Sullivan, & Preuschhoff, 2009)). In science the content

---

[2]There is a difference to the approach used in OECD PISA where the assessment is set up by experts independent of countries' curricula. In that sense, PISA is normative and defines a learning goal that should be reached by students at the age of 15 to be successful in their economic life (OECD, 1999).

domains assessed are life science, physical science, and earth science in grade four and biology, chemistry, physics, and earth science in grade eight; again the assessed cognitive domains are knowing, applying, and reasoning.

The test items in all TIMSS cycles are distinguished in multiple-choice items and open-ended items, the latter asking students to answer in an open format. Scorers assigned a number of score points to each answer which could be zero, one, and for some items even up to three score points. In the 2011 cycle of TIMSS, 217 mathematics and 217 science items were used in the grade eight assessments. Again, the items were grouped into blocks and the blocks were assembled to booklets using a spiral rotating design. Each student was assigned one test booklet with two testing session of about 40 min of testing time each.

The assessments – including stimulus material and items – were translated into the language(s) of instruction for each country. A thorough translation verification control took place to ensure that the meaning of the item and the difficulty in particular was not changed in the translation process. Each student was assessed in the language of instruction. In some countries, more than one target language was used; in Canada, for example, English and French were used in the assessment (and also for gathering the background information).

TIMSS Questionnaires

Besides the assessment component, TIMSS also administered several background instruments. For each participating country, a questionnaire had to be completed that included questions regarding the curriculum – especially the scope and content, its organization, monitoring, and evaluation systems as well as curricular materials and support (Mullis et al., 2009, pp. 82–83). The results of these questionnaires were published in encyclopedias (e.g., for TIMSS 2007: (Mullis & Martin, 2008)).

Furthermore, all sampled schools received a questionnaire to be completed by the school principal. This questionnaire included questions about the school demographics, the school organization, the goals of the school, the roles of the principal, school resources, the school climate, the parental involvement, the teacher recruitment, and about the teacher evaluation (Mullis et al., 2009, p. 84; 87).

All mathematics and science teachers of the sampled students received a questionnaire about their academic preparation and certification, the assignment, teacher induction, professional development, teacher characteristics, curriculum topics taught, class size, instructional time and activities, assessment in the class and homework assigned, computer and intranet use, calculator use, and on the emphasis on investigation (Mullis et al., 2009, p. 88; 93). Since mainly mathematics classes were sampled, in countries where students are taught in course systems, different science teachers could be responsible for different subgroups of the students. In countries where science is taught as separate subjects – physics, chemistry, biology, and earth science or any combination of those – each of the different science teachers had to complete one questionnaire.

In some countries, students also had more than one mathematics teacher responsible for different topics that were taught at the same grade – for example, algebra and analysis or algebra and geometry. It also occurred that some students had extra mathematics or science lessons – either extra tuition for lower performing students or special courses for high-performing students. Due to this fact, the number of mathematics and science teachers varies between but also within participating countries, and there is not necessarily a one to one match between students and mathematics and science teachers. This aspect must be considered when analyzing the TIMSS teacher data.

Another important factor regarding the TIMSS teacher data is that although the school and student data consists of a representative sample of schools and students of each participating country, the teacher data is not a representative sample. "The teachers in the TIMSS 2007 international database do not constitute a representative sample of teachers in the participating countries. Rather, they are the teachers of nationally representative samples of students. Therefore, analyses with the teacher data should be made with the students as the units of analysis and reported in terms of students who are taught by teachers with a particular attribute" (Foy & Olson, 2009, p. 11).

Finally all students who took part in the assessment received a questionnaire to be completed. The student background questionnaire mainly included information about the home background, such as "… number of books at home, availability of study desk, the presence of a computer, the educational level of the parents, and the extent to which students speak the language of instruction…" (Mullis et al., 2009, p. 94) but also questions about the lessons, the students' learning habits, and their attitudes towards the school in general but especially towards mathematics and science.

The TIMSS Sample

The aim of TIMSS is to analyze what happens in schools and within classes. Consequently, TIMSS uses grade-based[3] samples and applies a stratified three-stage cluster sampling approach where students were nested in classes and classes nested in schools – and in some countries even schools in regions.

---

[3] In comparison, the OECD PISA study applies an age-based sample and samples 15-year-old students in each participating country independent of the grade or even ISCED level that the students are enrolled in. For comparing school level factors on achievement, the grade-based approach seems to be beneficial since the grade distribution of groups of students with different achievement levels differs significantly in some countries. This is probably caused by the policies in some countries of having students that do not reach a certain ability level repeat a grade. For the immigrant population, another effect might cause that students are enrolled in grades lower than the median grade level in the countries, which is the change from one educational system into another with additional potential language problems. This problem will be discussed in more detail in Chapter 4B.

Prior to sampling, the population was also stratified. There are two different types of stratification: explicit stratification and implicit stratification. Explicit stratification means that the population is split into different groups based on its membership to the stratifying variables and then the sampling is executed separately for each group. If there is, for example, an interest in comparing rural and urban schools, in later analysis one could explicitly stratify by urbanization and sample the same number of rural and urban schools. Given that the variance of the variable of interest in both groups is similar, this would result in a relatively smaller sampling error of the difference between the two groups compared with a simple random sample on the expense that the overall sampling would be increased.

Implicit stratification means that the populations are ordered by the characteristics of the implicit stratification variables before a sample is drawn. For example, schools could be ordered by jurisdictions – like states – within a country. Then, with a systematic sampling approach (see below), all jurisdictions would be represented proportionally to their size. The advantage of this approach is that one can be sure that the random sample adequately represents the characteristics of the population with respect to the implicit stratum variable. In TIMSS, both implicit and explicit stratifications are used. The stratification variable for each country is listed in appendix in Foy and Olson (2009).

Within each stratum, the schools were ordered by the measure of size (MOS) which was usually the number of students either in the target grade (if this figure was available) or the number of students in the school or, where both were not available, any other measure of the size of the school. For the school sampling, a systematic sampling approach was chosen. This means that from the (stratified) list of schools, the first school was chosen randomly and then every nth school was selected where n is the sampling interval. The sampling interval is calculated as the MOS divided by the intended sample size. If, for example, the sampling interval was 100, then the school with the hundredth student after the first school was selected as the second school, the school with the two-hundredth student as the third school, and so on. This resulted in schools being sampled proportional to their size (PPS). Within the sampled schools, the classes were sampled randomly.

The aim was to sample 150 schools per country. Some countries selected more schools (up to 413 in Canada in TIMSS 1995) and some small countries had to select fewer schools. In Cyprus only 55 schools and in Kuwait 69 schools were sampled in TIMSS 1995 (Martin & Kelly, 1997a; Tables 2.16 and 2.17). Then one classroom was sampled randomly in each school for each grade. In TIMSS 1995, some countries chose to sample two classes per school, for example, Australia, Sweden, or the United States.[4] Then all students in the selected class were tested in

---

[4]This has become more common in later cycles of TIMSS since more countries were aware of the streaming that takes place within schools. This means that students are grouped in different classes by ability levels of the students. This results in significantly higher achieving and lower achieving classes in the schools concerned. To reduce variances between the sampled schools because of different ability classes being selected and to disentangle school and class effects, an increasing number of participating countries have chosen to select more than one class per school.

most countries. For example, in TIMSS 1995, in the Philippines 32 students were subsampled randomly in each sampled class because of very large classes.

If a school did not participate, the next school on the list of schools was selected as a replacement school. Since the replacement school is in the same explicit stratum and probably also in the same implicit stratum and has a comparable size, it is assumed that the replacement school is similar to the originally sampled school. If, however, a class or a student did not participate, no replacement was drawn since it was assumed that this could bias the sample. TIMSS applied rigorous standards for acceptable levels of non-response to avoid serious bias.

Due to the stratification and the sampling proportional to the size, weights were calculated at school level, at class level, and at within-class level (which is usually the same since all students in a class were selected). The weight is the inverse of the sampling probability. Consequently, large schools have a smaller weight due to the sample being proportional to size. But since larger schools usually have more classes, each class had a lower sampling probability. Thus, the sampled class within the school had a larger sampling probability. Also weight adjustments were calculated at school and at student level. This means that for each non-replaced nonparticipating school, the weight of the participating schools within the stratum was increased. The weights of the students within a sampled class were also increased if there were nonparticipating students in the class. The total student weight is the product of the school weight, the class weights, the student weight, the school adjustment factor, and the student adjustment factor. The student weights are summing up to the total student population in each country. For details about the sampling, the reader is referred to Chapter 5 in Foy and Olson (2009) and for the weighting to Foy and Olson (2009) Chapter 9.

The TIMSS Scores

In classical test theory (CTT), it is assumed that examinees have a certain ability or latent trait to be measured with a test. The estimate is specific to the particular test or set of items administered to the examinee. The measurement includes some errors; therefore, we are getting only an estimate of the examinees' actual ability. CTT offers different methods to estimate measurement error terms, the appropriateness of items in a test, and how well the test measures the ability. In classical test theory, examinees can be ranked by their ability and items can be ranked by their difficulty, but comparisons across different test forms administered to different populations are challenging.

In contrast to CTT, item response theory (IRT) estimates item properties – as item difficulty – that are expressed not in terms of percentages of correct answers achieved by a given population but rather in terms of the log of the odds of a person with a certain ability to achieve a particular response to the item. This means that the difficulty of the item is independent of the examinees tested. IRT is a measurement theory that estimates the examinees' probability of answering an item – given certain item characteristics – correctly. We thus want to know the probability for each examinee to answer each item correctly. In mathematical terms, this is described by P(x=1).

The simplest IRT model is one that considers only differences in the item difficulty. If I define the difficulty of an item I as b_i and the ability of an examinee v as θv, I come to the equation

$$P\left(x_{vi} = 1\right) = e^{\theta v - bi} / 1 + e^{\theta v - bi}$$

This is known as the RASCH or one-parameter model and was originally developed by the Danish mathematician Georg Rasch (Rasch, 1960).

This model has been extended by psychometricians by including a discrimination or slope parameter. The idea is that different items discriminate better or worse between more and less able students. This can be modeled by the formula

$$P\left(x_{vi} = 1\right) = e^{ai(\theta v - bi)} / \left(1 + e^{ai(\theta v - bi)}\right)$$

Especially for multiple-choice questions, respondents have a positive probability for guessing the correct response – even if they do not know the correct response. Examinees who do not know the answer have at least a chance of getting the item correct by marking randomly one of the response options. In IRT this can be modeled by introducing a guessing parameter into the model. To reflect this, a value c_i which indicates the probability of guessing the item i correctly is introduced which results in the formula

$$P\left(x_{vi} = 1\right) = c_i + \left(1 - c_i\right) e^{ai(\theta v - bi)} / 1 + e^{ai(\theta v - bi)}$$

There are also further extensions to these models, for example, for items with multiple score points (partial credit models – see (Masters, 1982)), or models for groups of examinees with a different response behavior. The methods for estimating the item parameters also differ significantly although all methods are iterative with more or less intense computational demands (Bock & Aitken, 1981). Further models refer to the examinees' scores which could also be modeled not as discrete values but as distributions of ability measures (Rubin, 2009; von Davier, Gonzalez, & Mislevy, 2009).

There is always, of course, the debate which model is the more appropriate, or even the best, or the only one to use. However, one should always examine if the method used is appropriate for the data to be analyzed. Especially the three-parameter models require a substantial amount of data.

It should also be noted that in theory some conditions must be met to allow the application of IRT models. The underlying latent trait must be unidimensional and the probability to get any item correct must not depend on any other item (local independence). However, different models were developed for tests that do not meet these criteria (see, e.g., (Brandt, 2008)).

In TIMSS IRT is used to analyze the test data and to assign achievement scores to students. Since the TIMSS assessment consists of different item types, different models are used for different items. A two-parameter IRT model is used for dichotomous constructed response items and a three-parameter model for dichotomous multiple-choice items. Since some constructed response items were not only coded to correct or incorrect, but partially correct answers were also assigned partially correct codes, using a generalized partial credit two-parameter model to describe them (Olson, Mullis, & Martin, 2008, Chapter 11.2)).

For the calculation of student proficiency scores, the data from the current and previous cycles were scaled together to achieve a common scale and consequently achievement scores for all TIMSS cycles on one common metric. This resulted in student achievement scores on a logit metric with values usually between -3 and 3 – and in extreme cases between -5 and 5, with lower values indicating lower ability levels of students and higher values higher ability levels of students.

As described above, in TIMSS a matrix-sampling design is applied. This means that tested students are assigned only a fraction of the available items. This design is less optimal for estimating the ability of individual students but is very efficient in giving good population proficiency estimates.[5] To account for the measurement error introduced by the matrix-sampling design and to achieve good estimates for the population – and subpopulation –, estimates for the plausible value are calculated.

Plausible values are drawn from the ability distribution of each student. To achieve this, the background data gathered in TIMSS is analyzed with a factor analysis, and as many factors are extracted, until 90 % of the variance in the background data was accounted for. From the multivariate normal distribution of the conditioned ability distribution of the students ($P(\Theta_j \mid y_j, \Gamma, \Sigma)$), five values are randomly drawn. These values are called plausible values and the variance between the plausible values is an estimator for the measurement error. For example, take a student who gives wrong answers to items that are characterized as easy but answers harder items correctly and who has a very different score compared to other students with the same background information, e.g., a boy scoring high on a reading scale on which girls usually score higher. The ability distribution for this student would be relatively wide compared with, e.g., a high-scoring girl that answered all easy items as well as some of the difficult items correctly. The girl's ability distribution would be much narrower. Consequently, the variance of the five randomly drawn plausible values would also be higher for the boy in this example (Olson et al., 2008, Chapters 11.2.3–11.2.5).

---

[5] This is also the reason why it is not recommended to use the TIMSS test to analyze individual students or to give feedback to individual students or very small groups of students (see, e.g., Chapter 3.2.1 in (Mirazchiyski, 2013)).

Analyzing TIMSS Data

When analyzing TIMSS data, the sampling weights must always be considered to ensure that the results are representative of the total population. The other effect of the sampling approach is that the effective sample size is smaller compared with a simple random sample of students. When using the total weight, statistical standard software overestimates the sample size and thus underestimates variances or standard errors. To overcome this challenge, TIMSS recommends the use of the Jackknifing procedure (JK2) which is a replication method where the statistics of interest are calculated several times using replication weights. The variance of the results is an estimator for the sampling error (see (Foy & Olson, 2009)).

The other challenge when analyzing TIMSS data is the correct usage of the plausible value scores. For mean statistics, each of the five plausible values can be used in the analysis. Usually the first plausible value is used or sometimes the results of the analysis of the five plausible values are averaged. The plausible values should never be averaged before the analysis.

As described above, the variance of the plausible values is an estimation of the measurement error. Therefore, the variance of the results of any analysis conducted five times with each of the plausible values separately is an estimation of the impact of the measurement error on the results.

The total error for a statistic based on TIMSS data that makes use of the plausible values is then the square root of the sum of the squared sampling error and the squared measurement error since sampling error and measurement error are assumed to be independent: $SE_{tot} = \sqrt{SE_{mea}2 + SE_{samp}^2}$.

Methods of Analysis

To assist researchers, IEA has developed an international database analyzer (IDB Analyzer) which considers the sampling method as well as the scoring methods when calculating statistics. The IDB analyzer is a plug-in for SPSS. In the analysis, the IDB analyzer is used to calculate the statistics and especially the standard errors of differences. The IDB analyzer considers the weights, the sampling strategy, and the plausible values correctly by applying the Jackknife procedure and by calculating each statistic five times with each plausible value separately, then calculating the variance of the results.

Significance and Relevance of Differences

With the help of the IDB analyzer, the correct standard errors of the statistics can be calculated. I follow the general approach used in the TIMSS international report and define two statistics as statistically significant difference at a 95 % confidence level.

This means that the probability of assuming a difference of two statistics in the population that we generalize on as significant – although it is not – is 95 %. In statistic terms, this means that the probability for a type I error is 5 % (for TIMSS, see Chapter 12.4.1 in (Martin, Mullis, & Chrostowski, 2004)). However, with a sufficiently large sample size and a test tool that delivers sufficiently accurate measures at individual level, all differences can become statistically significant. This leads to one of the critics towards results from large-scale assessment studies – namely, that all differences become statistically significant.

The question then is if the differences are meaningful, is a difference of five score points in mathematics achievement between boys and girls in TIMSS 2007 in Slovenia relevant? In the interpretation of the results from the quantitative analysis, we need to examine if differences are statistically significant because nonsignificant differences are definitely not meaningful. But I also want to interpret the results by referring to results from TIMSS 1995.

In TIMSS 1995, countries tested students not only in one target grade but in two adjacent grades. This resulted in testing students in grades three and four and in grades seven and eight. The assessments of adjacent grades were made with the same instruments, and the scores were calculated on the same metric with a mean of 500 and a standard deviation of 100 across both grades. Since I am analyzing the grade eight data in this publication, I focus on the differences between the grade seven and grade eight results in TIMSS 1995 to evaluate the learning effect. Table 3.1 displays the differences between these grades for each participating country in TIMSS 1995.

The differences between the two grades differ between the countries. As Beaton et al. concluded: "Increases in mean performance between the two grades ranged from a high of 49 points in Lithuania to a low of 8 points in the Flemish-speaking part of Belgium and 7 points in South Africa. This degree of increase can be compared to the difference of nearly 30 points between the international average of 513 at eighth grade and that of 484 at seventh grade" (Beaton, Mullis, et al., 1996, p. 27). Although the differences are varying between countries and probably also in subgroups of students as well as across time, we might use the 30-point score difference to give some meaning to differences within grade eight groups of students. When interpreting a difference of 25 score points between first-generation immigrant students and native students in Latvia in 1995 – as we will see in Chap. 4A – we might interpret this as equaling nearly the amount of what students learn on average in one school year. This interpretation should be made with great care and reservation, but it might help judge if a difference is not only statistically significant but also meaningful.

In all tables in this publication, the significance of all difference was determined by dividing the difference by its standardized error and comparing it to the critical value of 1.96. This means that the differences are statistically significantly different on a 95 % confidence level ($\alpha = 0.05$) based on a two-tailed significance test with infinite degrees of freedom.

Table 3.1 Achievement differences in mathematics between lower and upper grades (seventh and eighth grades)

Country	Seventh Grade Mean	Eighth Grade Mean	Eighth-Seventh Difference	
[1] Lithuania	428 (3.2)	477 (3.5)	49 (4.7)	
France	492 (3.1)	538 (2.9)	46 (4.3)	
Norway	461 (2.8)	503 (2.2)	43 (3.6)	
Singapore	601 (6.3)	643 (4.9)	42 (8.0)	
Sweden	477 (2.5)	519 (3.0)	41 (3.9)	
Czech Republic	523 (4.9)	564 (4.9)	40 (7.0)	
[1] Switzerland	506 (2.3)	545 (2.8)	40 (3.6)	
Spain	448 (2.2)	487 (2.0)	39 (3.0)	
Slovak Republic	508 (3.4)	547 (3.3)	39 (4.7)	
New Zealand	472 (3.8)	508 (4.5)	36 (5.9)	
[†] Scotland	463 (3.7)	498 (5.5)	36 (6.6)	
Hungary	502 (3.7)	537 (3.2)	35 (4.9)	
Russian Federation	501 (4.0)	535 (5.3)	35 (6.6)	
Japan	571 (1.9)	605 (1.9)	34 (2.7)	
Canada	494 (2.2)	527 (2.4)	33 (3.3)	
[1] Latvia (LSS)	462 (2.8)	493 (3.1)	32 (4.2)	
Portugal	423 (2.2)	454 (2.5)	31 (3.3)	
Korea	577 (2.5)	607 (2.4)	30 (3.5)	
[†2] England	476 (3.7)	506 (2.6)	30 (4.5)	
Cyprus	446 (1.9)	474 (1.9)	28 (2.7)	
Ireland	500 (4.1)	527 (5.1)	28 (6.6)	
Iran, Islamic Rep.	401 (2.0)	428 (2.2)	27 (2.9)	
Iceland	459 (2.6)	487 (4.5)	27 (5.2)	
Hong Kong	564 (7.8)	588 (6.5)	24 (10.2)	
[†] United States	476 (5.5)	500 (4.6)	24 (7.2)	
[†] Belgium (Fr)	507 (3.5)	526 (3.4)	19 (4.9)	
[†] Belgium (Fl)	558 (3.5)	565 (5.7)	8 (6.7)	
Countries Not Satisfying Guidelines for Sample Participation Rates (See Appendix A for Details):				
Australia	498 (3.8)	530 (4.0)	32 (5.5)	
Austria	509 (3.0)	539 (3.0)	30 (4.3)	
Bulgaria	514 (7.5)	540 (6.3)	26 (9.8)	
Netherlands	516 (4.1)	541 (6.7)	25 (7.8)	
Countries Not Meeting Age/Grade Specifications (High Percentage of Older Students: See Appendix A for Details):				
Slovenia	498 (3.0)	541 (3.1)	43 (4.3)	
Romania	454 (3.4)	482 (4.0)	27 (5.3)	
[†1] Germany	484 (4.1)	509 (4.5)	25 (6.1)	
Colombia	369 (2.7)	385 (3.4)	16 (4.4)	
Countries With Unapproved Sampling Procedures at Classroom Level (See Appendix A for Details):				
Denmark	465 (2.1)	502 (2.8)	37 (3.5)	
Greece	440 (2.8)	484 (3.1)	44 (4.2)	
South Africa	348 (3.8)	354 (4.4)	7 (5.9)	
Thailand	495 (4.8)	522 (5.7)	28 (7.5)	

-10 0 10 20 30 40 50 60
±2 SE of the
Difference

↑
Difference

Source: TIMSS 1995 Assessment. Copyright © 1996 International Association for the Evaluation of Educational Achievement (IEA). Publisher: TIMSS & PIRLS International Study Center, Lynch School of Education, Boston College

Focus on Mathematics

When analyzing the educational outcome for immigrants and native students, a main focus will be the achievement measured in TIMSS. As stated above, TIMSS measured mathematics and science achievement, while PIRLS measures reading literacy abilities. Clearly, mathematics, science, and reading achievement is only one aspect of educational outcome. Although I will also focus on attitudes towards school and mathematics as well as on self-esteem with respect to mathematics, it is clear that this is only a limited focus.

I do not investigate the creativity of the students, their social competencies, their cooperation, their health and well-being, and many other aspects that are and should be foci of education. This is clearly a limitation of the study at hand, but the limitation was necessary to make the in-depth analysis feasible. And one can also argue that although literacy, mathematics, and science abilities of students are not exhaustive learning outcomes, they are nevertheless basic skills. The ability to be a good and efficient learner in these subjects also shows the ability of students to learn other subjects. In later life, there are strong relationships between these basic skills and other aspects in life, as shown by the OECD PIAAC survey which assessed various skills of adult populations and their relationship to other aspects of life: "The Survey of Adult Skills collected information on four dimensions of well-being: the level of trust in others, political efficacy or the sense of influence on the political process; participation in associative, religious, political or charity activities (volunteering); and self-assessed health status. Over all, literacy proficiency has a positive relationship with all four of the outcomes considered, net of the effects of education, socio-economic background, age, gender, and immigrant background" (OECD, 2013a, p. 234).

Although school education does and should do more than preparing people for work, being employed as an adult is surely also a positive outcome of education – not only in economic aspects but also in a more holistic view. In this respect, the PIAAC survey revealed that: "An individual who scores one standard deviation higher than another on the literacy scale (around 46 score points) is 20 % more likely to participate in the labour market, ..." (OECD, 2013a, p. 227).

With students, the mathematics, science, and reading literacy achievement is strongly correlated. As described above, in TIMSS/PIRLS 2011 countries were offered the opportunity to assess the same students in grade four. For the countries that made use of this option, Table 3.2 shows the correlations at student level among these three scales. Table 3.2 shows the correlation between reading literacy and mathematics, reading literacy and science, and mathematics and science achievement. The difference of the correlations between reading literacy and mathematics, reading literacy and science, and their significance is displayed.

As can be seen from Table 3.2, all three scales correlate quite strongly. Internationally, reading literacy and mathematics scores correlate to 0.78, reading literacy and science correlate to 0.85, and mathematics and science correlate to

Table 3.2 Correlation of reading, mathematics, and science scores in TIMSS 2011 grade four

Country	Reading and math		Difference R-M vs R-S		Reading and science		Math and science	
	Corr	SE	Diff	sign	Corr	SE	Corr	SE
Saudi Arabia	0.72	0.02	0.12	⇧	0.84	0.01	0.80	0.02
Canada (Quebec)	0.67	0.01	0.12	⇧	0.79	0.01	0.75	0.01
Norway	0.74	0.02	0.11	⇧	0.85	0.01	0.79	0.02
Sweden	0.73	0.01	0.11	⇧	0.84	0.01	0.78	0.01
Spain	0.72	0.01	0.10	⇧	0.82	0.01	0.79	0.01
Malta	0.77	0.01	0.09	⇧	0.86	0.00	0.81	0.01
Portugal	0.76	0.01	0.09	⇧	0.85	0.01	0.85	0.01
Australia	0.81	0.01	0.09	⇧	0.90	0.01	0.86	0.01
Finland	0.72	0.02	0.09	⇧	0.81	0.01	0.80	0.01
Italy	0.75	0.01	0.09	⇧	0.84	0.01	0.80	0.01
Austria	0.78	0.01	0.08	⇧	0.87	0.01	0.82	0.01
Germany	0.79	0.01	0.08	⇧	0.87	0.01	0.81	0.01
Honduras, Republic of	0.74	0.02	0.08	⇧	0.83	0.01	0.84	0.01
Romania	0.79	0.02	0.08	⇧	0.87	0.01	0.86	0.01
Slovenia	0.81	0.01	0.08	⇧	0.89	0.01	0.87	0.01
Croatia	0.76	0.01	0.08	⇧	0.84	0.01	0.80	0.01
Russian Federation	0.73	0.01	0.08	⇧	0.81	0.01	0.83	0.01
Hong Kong, SAR	0.69	0.01	0.08	⇧	0.77	0.01	0.77	0.01
Botswana	0.84	0.01	0.07	⇧	0.91	0.00	0.90	0.01
Georgia	0.76	0.01	0.07	⇧	0.83	0.01	0.85	0.01
Slovak Republic	0.81	0.02	0.07	⇧	0.88	0.01	0.88	0.01
Singapore	0.84	0.01	0.07	⇧	0.91	0.00	0.86	0.01
Ireland	0.79	0.01	0.07	⇧	0.86	0.01	0.84	0.01
Iran, Islamic Republic of	0.82	0.01	0.07	⇧	0.89	0.01	0.87	0.01
Poland	0.82	0.01	0.06	⇧	0.88	0.01	0.85	0.01
Hungary	0.84	0.01	0.06	⇧	0.90	0.01	0.88	0.01
United Arab Emirates	0.84	0.01	0.06	⇧	0.90	0.00	0.87	0.00
United Arab Emirates (Abu Dhabi)	0.84	0.01	0.06	⇧	0.90	0.01	0.86	0.01
United Arab Emirates (Dubai)	0.85	0.01	0.06	⇧	0.91	0.00	0.88	0.00
Czech Republic	0.79	0.01	0.06	⇧	0.84	0.01	0.82	0.01
Chinese Taipei	0.78	0.01	0.05	⇧	0.84	0.01	0.81	0.01
Northern Ireland	0.82	0.01	0.04	⇧	0.86	0.01	0.84	0.01
Azerbaijan, Republic of	0.59	0.02	0.04		0.63	0.02	0.76	0.01
Lithuania	0.82	0.01	0.04	⇧	0.85	0.01	0.86	0.01
Oman	0.81	0.01	0.03	⇧	0.84	0.01	0.88	0.01
Qatar	0.85	0.01	0.02	⇧	0.87	0.01	0.87	0.01
Morocco	0.69	0.01	0.01		0.71	0.01	0.81	0.01
Int. Avg.	0.78	0.00	0.07	⇧	0.85	0.00	0.83	0.00

0.83. I conclude that in general and across all countries, students doing well in one of the subjects are usually also doing quite well in the other subjects. Consequently, the results I found analyzing one of the scales have a high potential to be valid for the other two subjects as well.

The reason why I focus on TIMSS instead of PIRLS data is that PIRLS measures the abilities of grade four students, but the background information gathered for grade four students, as I will show in Chap. 4A, is less reliable especially with regard to the immigrant status. TIMSS data, on the other hand, allows me to analyze the mathematics and science abilities of the students as achievement outcomes. Therefore, Chap. 4A shows the results for mathematics and science for immigrant and native students. From Chap. 4B on, I will focus on the mathematics achievement of the students. As stated above, the results found for mathematics have a high potential to be valid also for science – and, to a certain extent, for reading literacy.

The focus on mathematics for most part of this research has two main reasons. One reason is the lower influence of language on the achievement results. Table 3.2 shows that the correlation between reading literacy abilities and mathematics is statistically significantly lower than for reading literacy abilities and science in all but two countries. As we will see in Chap. 4B, a high proportion of immigrant students is coming with a different language background, and their lack in language proficiency thus impacts their learning and achievement. Therefore, examining mathematics achievement is less disadvantaging for immigrant students than investigating their science achievement.[6]

The second reason is based on the TIMSS design. In TIMSS, mathematics classes – one or more – were sampled randomly in each sampled school because in most countries mathematics classes were identified as exhaustive and mutually exclusive groups of students. The mathematics teachers of the sampled students also completed a questionnaire and the data is used in the analysis presented, too.

For science, the situation is more complicated. First, in some countries science is taught as one subject, whereas in other countries, science is separated into different subjects – physics, biology, chemistry, earth science, and others. This is the case in, for example, Germany, Austria, or Slovenia. Moreover in some countries there is a different emphasis on the different science domains in different school years, although all are taught in one subject called science. Therefore science is taught by different teachers in different grades in the participating countries. This makes it difficult to obtain reliable information from the current teachers in any grade about all science subjects.

Furthermore, as also described above, in several countries, students are taking more than one science course per grade, which was probably the main reason for sampling mathematics classes rather than science classes. This results in sampled students being linked to more than one science teacher in some countries. In other countries, some but not all students from several science were included in the international database which results in small subsamples of a number of science classes

---

[6]One caveat here is that the correlations shown in Table 3.2 are based on grade four student data, whereas it is data from grade four students which is mainly analyzed in this study.

per school. This makes an analysis of science classes with TIMSS data very difficult.

For the analysis of science teacher data, an additional obstacle is that in cases where there is more than one science teacher, the different impacts from the different teachers can hardly be disentangled. For example, if I assess the influence of male or female mathematics teachers on students in science, I have to distinguish between students being taught by only male science teachers, only female science teachers, or a combination of male and female science teachers – and again the influence of each is hard to disentangle.

Chapter 4A
Immigrant Students in TIMSS

Abstract In this chapter, the trends in percentages and the achievement trends in both mathematics and science for immigrant students are analyzed using data from the 1995, 2003, and 2007 cycles of TIMSS. The aim is to gain more insights as a basis for more in-depth analysis of TIMSS 2007 data. First, trends in percentages of first-generation immigrant students in grade eight and then in grade four are analyzed, followed by an analysis of trends of second-generation immigrant students in grade eight. Then, achievement differences between immigrant and native students will be explored. The data reveal that overall, immigrant students – especially first-generation immigrant students – are outperformed, and increasingly so, by native students in both mathematics and science. The first-generation immigrant population has increased between 1995 and 2007 in a large number of countries.

Keywords Achievement trends • Mathematics achievement • Science achievement

I have so far given an introduction into the topic and outlined the research questions that guide this publication. A literature review on the topics that will be analyzed starting from this chapter was conducted and the following analyses will build on these findings. The previous chapter has introduced the data and methods that will be used in this and the following three chapters.

In this chapter, the trends in percentages and the achievement trends in mathematics and science for immigrant students are analyzed. The aim is to get an overview before more in-depth analysis of TIMSS 2007 data will be performed.

Trends in Percentages of First-Generation Immigrant Students in Grade Eight

The percentages of students with an immigrant background enrolled in schools vary substantially between countries – and in most countries also within the country. First I will analyze the percentages of first-generation immigrant students in the

33 ▆▆

A. Were you born in <country>?

Fig. 4A.1 TIMSS 2007 student questionnaire, variable BS4GBORN (SOURCE: TIMSS 2007 Assessment. Copyright © 2009 International Association for the Evaluation of Educational Achievement (IEA). Publisher: TIMSS & PIRLS International Study Center, Lynch School of Education, Boston College)

school systems. The focus is on grade eight students. In the analysis TIMSS grade eight data is used.

In the TIMSS cycles 1995, 1999, 2003, and 2007, the students were asked if they were born in the country of residence. For example, in TIMSS 2007 the following question was included at the end of the student questionnaire (Fig. 4A.1):

The participating countries had to adapt this international version of the question by replacing "<country>" with their country name. Table 4A.1 shows the percentage of students in grade eight for each of the TIMSS cycles who reported not to be born in the country of residence for each of the country participating in at least one of the TIMSS cycles.

As can be seen from Table 4A.1, out of 22 countries that participated in TIMSS 1995 and TIMSS 1999, in two countries the percentage of immigrant students decreased statistically significantly (Iran and the Philippines), in four countries the percentage increased (Hong Kong, Latvia, Slovak Republic, and Thailand), and in the other countries the percentages did not change statistically significantly.

When comparing the TIMSS 1999 results with the TIMSS 2003 results, I observe that out of 32 countries that participated in both cycles, there is a decrease in one country (Tunisia) but an increase in 21 countries. The trend between TIMSS 2003 and TIMSS 2007 shows a decrease in ten countries and an increase in twelve countries.

Obviously, the percentage of first-generation immigrant students in the educational system increased in a large number of countries with the major increase between 1999 and 2003. In some of the countries, the increase is quite significant. For example, the percentage of first-generation immigrants in South Africa increased from 13 to 33 % between 1999 and 2003. In Saudi Arabia, we see an increase from 9 to 17 % between 2003 and 2007. In Lebanon we observe an increase from 12 to 21 % between 2003 and 2007 and in Kuwait from 7 to 16 % between 1995 and 2007. Also Bulgaria showed an increase between 2003 and 2007 from 5 to 10 %. Of course this data refers to the student responses in TIMSS and is subject to response errors. Consequently, these figures do not always match the official statistics. But even if one takes into account an error margin, the general increases are evident and they pose a challenge to the educational systems of the concerned countries. So, whatever perspective is taken, the conclusion is that the percentage of immigrant students increased overall for grade eight students based on TIMSS.

Table 4A.1 Trends in percentages of first-generation immigrant students in TIMSS grade eight

| Country | 1st generation immigrants | | | | | | | | trends | | | | | | | |
| | 1995 | | 1999 | | 2003 | | 2007 | | 95->99 | | 99->03 | | 03->07 | | 95->07 | |
	Perc.	SE	Perc.	SE	Perc.	SE	Perc.	SE	%	sign	%	sign	%	sign	%	sign
Armenia	11	0.8			4	0.5	11	0.9					7	⇦		
Australia	6	0.5	14	1.3	15	1.3	11	0.9	3	⇑	1	⇑	−4	⇨	0	⇑
Austria																
Bahrain					10	0.5	14	0.7					4	⇦		
Belgium (Flemish)	3	0.5	4	0.9	7	0.8			1	⇑	3	⇦				
Belgium (French)	9	0.9														
Bosnia and Herzegovina							24	1.2								
Botswana			3	1.0	4	0.8	5	0.4					1	⇑		
Bulgaria					5	0.6	10	0.7			2	⇦	5	⇦		
Canada (British Columbia)			10	0.7			16	1.3								
Canada (Ontario)					16	1.6	13	1.3			2		−3	⇑		
Canada (Quebec)					6	1.1	8	0.7			1		1	⇑		
Chile					5	0.4	6	0.4				⇦				
Chinese Taipei			3	0.3	2	0.3	5	0.5				⇦	4	⇦		
Colombia	3	0.3	1	0.1											2	⇦
Cyprus	10	0.4	8	0.4	12	0.5	10	0.5	−1	⇑	4	⇦	−2	⇨	1	⇑
Czech Republic	1	0.2	1	0.3			3	0.3	0	⇑					1	⇦
Denmark	6	0.5														
Egypt					26	1.8	43	1.9					17	⇦		
El Salvador							5	0.5								

(continued)

Table 4A.1 (continued)

| Country | 1st generation immigrants | | | | | | | | trends | | | | | | | |
| | 1995 | | 1999 | | 2003 | | 2007 | | 95->99 | | 99->03 | | 03->07 | | 95->07 | |
	Perc.	SE	Perc.	SE	Perc.	SE	Perc.	SE	%	sign	%	sign	%	sign	%	sign
England	5	0.5	5	0.6	7	0.9	8	0.7	0	⇑	1	⇑	1	⇑	3	⇐
Estonia					4	0.4										
Finland			3	0.5												
France																
Georgia							6	0.6								
Germany	10	0.8														
Ghana					31	1.8	18	1.4					-13	⇒		
Greece	6	0.4														
Hong Kong, SAR	13	0.9	18	1.2	27	1.0	26	1.3	5	⇐	9	⇐	-1	⇑	13	⇐
Hungary	2	0.2	2	0.3	3	0.3	3	0.4	0	⇑	1	⇑	0	⇑	1	⇐
Iceland	7	0.5														
Indonesia			1	0.2	13	1.0	16	1.2			12	⇐	3	⇐		
Iran, Islamic Republic of	2	0.3	1	0.2	1	0.2	1	0.1	-1	⇒	1	⇐	-1	⇒	-1	⇒
Ireland	3	0.3														
Israel	14	2.1	16	1.7	14	1.0	13	1.2	2	⇑	-3	⇑	-1	⇑	-1	⇑
Italy			3	0.3	5	0.4	5	0.4			2	⇐	0	⇑		
Japan			1	0.1	1	0.2	1	0.2			0	⇐	0	⇑		
Jordan			16	0.9	25	1.3	13	1.3			8	⇐	-11	⇒		
Korea, Republic of			2	0.2	1	0.2	0	0.1			0	⇑	-1	⇒		
Kuwait	7	0.8					16	0.8							9	⇐
Latvia	1	0.2	7	1.8	4	1.4			6	⇐	-3	⇑				
Lebanon					12	1.4	21	1.4					8	⇐		

Country	%		%		%		%		%		Δ		Δ		Δ		Δ	
Lithuania	2	0.3	2	0.3	2	0.3	4	0.3	5	0.4	0	⇧	2	⇦	0	⇧	3	⇦
Macedonia			3	0.3			8	0.8	8	0.7			5	⇦	3	⇦		
Malaysia			2	0.2			5	0.6	7	0.4			3	⇦				
Malta							13	1.0	7	0.4			4	⇦				
Moldova			9	1.1			13	1.0	14	0.8								
Mongolia																		
Morocco	5	0.5	6	0.7			12	1.3	7	0.7			6	⇦	−5	⇨	1	⇨
Netherlands	12	0.6	7	1.0			7	0.6			2	⇧	0	⇧				
New Zealand	5	0.4	14	0.8			14	1.3			1	⇧	1	⇧				
Norway							8	0.6	6	0.4					−1	⇨	1	⇦
Oman									15	1.2					−3	⇧		
Palestinian National Authority	2	0.1	21	1.4			24	1.2	21	1.2	19	⇦	−18	⇨				
Philippines							3	0.3										
Portugal	8	0.7																
Qatar									26	0.5								
Romania	4	0.6	3	1.0			2	0.2	4	0.5	−1	⇧	−1	⇧	2	⇦	0	⇧
Russian Federation	6	0.8	7	0.7			9	0.6	8	0.7	0	⇧	2	⇦	−1	⇦	2	⇧
Saudi Arabia							9	0.7	17	1.1					9	⇦		
Scotland	8	0.5					5	0.4	6	0.6					1	⇧	−1	⇧
Serbia							10	0.8	7	0.5					−3	⇨		
Singapore	8	0.4	9	0.6			13	0.6	11	0.5	1	⇧	4	⇦	−2	⇨	3	⇦
Slovak Republic	2	0.2	2	0.3			3	0.4			1	⇦	1	⇦				
Slovenia	3	0.3	4	0.4			4	0.4	5	0.4	0	⇧	0	⇧	1	⇧	2	⇦
South Africa	13	1.1	12	1.0			33	1.7			−1	⇧	20	⇦			2	⇦
Spain	3	0.3																

(continued)

Table 4A.1 (continued)

| Country | 1st generation immigrants | | | | | | | | trends | | | | | | | |
| | 1995 | | 1999 | | 2003 | | 2007 | | 95->99 | | 99->03 | | 03->07 | | 95->07 | |
	Perc.	SE	Perc.	SE	Perc.	SE	Perc.	SE	%	sign	%	sign	%	sign	%	sign
Spain (Basque Country)					3	0.5	6	0.7					3	⇦		
Sweden	8	0.6			9	0.9	8	0.6					-2	⇧	0	⇧
Switzerland	12	0.6														
Syria, Arab Republic of							24	1.1								
Thailand	1	0.2	4	0.6	0	0.1	1	0.1	3	⇧					-1	⇨
Tunisia			5	0.7			5	0.4			-4	⇨	4	⇦		
Turkey			4	0.4			1	0.2								
Ukraine							7	0.6								
United Arab Emirates (Dubai)							52	1.4								
United States	7	0.6	8	0.8	8	0.5	10	0.6	1	⇧	1	⇧	1	⇧	2	⇦
United States (Indiana)					5	0.7										
United States (Massachusetts)							9	0.9								
United States (Minnesota)							7	1.8								

Trends in Percentages of First-Generation Immigrant Students in Grade Four

The fourth grade students in TIMSS as well as in PIRLS were asked the same question regarding their immigration status. TIMSS assessed fourth grade students in 1995, 2003, and 2007 and PIRLS assessed fourth grade students in 2001 and in 2006. It would be interesting to examine the reading achievement of immigrant students which was only measured for grade four students in PIRLS and also the achievement differences and other factors for immigrant students at the end of primary education. The problem is, however, that the information on the immigrant status is not very reliable for grade four students. Table 4A.2 shows the percentages of first-generation immigrants as reported by grade four students for all countries that participated in more than one of the assessments.

The percentages vary quite substantially even in cycles where the measurement was made almost at the same time. We can also see that the figures go up and down considerably between the different points in time. If we focus only on the percentages of immigrants in PIRLS 2006 and TIMSS 2007 – assessments that took place in two adjacent years – we see, for example, in Austria 6 % of immigrants in 2006 reported by PIRLS grade four students and 16 % in 2007 reported by TIMSS grade four students. Although the percentages might have increased, it is very unlikely that such a dramatic change occurred.

The largest difference between the percentages of immigrant students based on PIRLS 2006 and TIMSS 2007 grade four students can be observed for Kuwait. In Kuwait 8 % of the grade four students in PIRLS answered to be immigrants in 2006, but 41 % of the grade four students in TIMSS 2007 indicated to be immigrants. In contrast to the quite stable estimates for the percentage of immigrants in grade eight, these percentages are neither in line with each other nor with the trends that we can observe for the grade eight students. Consequently, the further analysis will focus only on the grade eight data since it proves to be a more solid base.

Trends in Percentages of Second-Generation Immigrant Grade Eight Students

The TIMSS data also allows for identifying second-generation immigrant students. The students were asked where their parents were born (Fig. 4A.2).

These questions were asked independently of the question about the students' place of birth. Again, the participating countries had to adapt this international version of the question by replacing "<country>" with their country name. Consequently, students could respond that their father or mother was not born in the country but that they were or that both of their parents were born in the country, but they (the students) were not. Another possible response was that they were not born in the country but that both of their parents were.

Table 4A.2 Trends in percentages of students not born in the country of schooling

Country	TIMSS 1995		PIRLS 2001		TIMSS 2003		PIRLS 2006		TIMSS 2007	
	pct	SE	pct	SE	pct	SE	pct	SE	pct	SE
Armenia					4	0.4			30	1.7
Australia	9	0.8			16	0.9			16	0.8
Austria							6	0.5	16	0.6
Belgium (Flemish)					7	0.7	5	0.4		
Bulgaria			7	0.7			1	0.2		
Canada (British Columbia)							12	1.0	19	1.0
Canada (Ontario)			22	1.4			10	1.1	22	1.9
Canada (Quebec)			21	1.2	7	0.7	8	0.8	14	1.2
Cyprus	12	0.6	14	0.8	12	0.7				
Czech Republic	2	0.2	5	0.5					4	0.4
Denmark							5	0.4	10	0.8
England			15	1.3	15	1.0	8	0.7	14	0.7
Georgia							2	0.3	14	1.2
Germany			21	0.8			5	0.5	10	0.5
Hong Kong, SAR	19	1.7	30	2.0	27	1.5	18	0.9	25	1.6
Hungary	4	0.3	11	0.8	7	0.6	2	0.3	7	0.7
Iceland	12	1.9	16	0.5			7	0.4		
Iran, Islamic Republic of	11	0.8	3	0.4	3	0.4	3	0.4	2	0.4
Israel	16	1.5	25	1.2			7	0.6		
Italy			5	0.4	4	0.3	5	0.4	5	0.4
Japan					2	0.2			1	0.2
Kuwait	13	1.0	26	0.9			8	0.6	41	1.7
Latvia	4	0.4	9	1.4	10	0.8	26	1.3	8	0.7
Lithuania			5	0.7	6	0.4	1	0.2	8	0.7
Moldova			40	3.8	20	1.3	6	0.6		
Morocco			6	0.9	26	2.4	5	0.6	12	1.2
Netherlands	11	1.0	8	0.7	7	0.8	4	0.4	18	1.0
New Zealand	10	0.7	18	1.0	16	0.9	14	0.8	26	0.9
Norway	5	0.4	9	0.7	9	0.6	4	0.4	5	0.4
Qatar							18	0.4	58	0.5
Romania			4	0.9			1	0.2		
Russian Federation			9	0.7	14	0.9	5	0.4	7	0.8
Scotland	10	0.8	39	4.4	17	1.0	4	0.4	13	0.8
Singapore	8	0.6	21	1.0	21	1.2	10	0.4	10	0.4
Slovak Republic			9	1.3			1	0.2	3	0.4
Slovenia	4	0.4	11	0.9	3	0.4	2	0.3	12	0.5
Sweden			12	0.9			5	0.5	11	0.8
Tunisia					0	0.1			12	1.2
United States	9	0.7	19	1.0	20	0.8	8	0.6	19	0.7
Yemen					56	3.0			44	2.3

32

A. Was your mother (or stepmother or female guardian) born in <country>?

B. Was your father (or stepfather or male guardian) born in <country>?

Fig. 4A.2 TIMSS 2007 student questionnaire, variables BS4GMBRN and BS4GFBRN (SOURCE: TIMSS 2007 Assessment. Copyright © 2009 International Association for the Evaluation of Educational Achievement (IEA). Publisher: TIMSS & PIRLS International Study Center, Lynch School of Education, Boston College)

There are different approaches on how to define second-generation immigrants. They mostly deviate in defining a child as a second-generation immigrant if the father, or the mother, or both parents were born outside the country of residence.

This publication follows the definition used by the OECD (2010c) and EC (Eurydice network, 2009), defining second-generation immigrants as students who were born in the country of residence, but at least one of their parents was not born in the country of residence. This seems to be the most commonly used definition. When applying this definition to TIMSS data, I receive the results displayed in Table 4A.3.

Table 4A.3 shows substantial changes in the number of second-generation immigrant students in several countries. Especially between 2003 and 2007, there were quite a number of countries with significant increases in the number of second-generation immigrant students in the school population, with statistically significant increases in Botswana, Cyprus, Hong Kong, Hungary, Japan, Norway, Scotland, Serbia, Basque Region of Spain, Syria, Tunisia, and the United States of America. Between 1999 and 2003, the number of second-generation immigrant students decreased rather than increased. This trend could be due to reduced immigration or to changing immigration policies that, e.g., do not permit spouses to accompany their partners. Another explanation might be that children of immigrants are not attending school in the immigrant country. This effect can only be analyzed by further in-depth research carried out separately for each country and based on supplementary information.

So, in summary, I cannot say that the percentage of second-generation immigrant students increased in all countries, but I find a decent number of countries where the percentage of second-generation immigrant students increased between the first cycle of TIMSS in 1995 and the TIMSS cycle of 2007.

Table 4A.3 Trends in percentages of second-generation immigrant students in TIMSS grade eight

Country	2nd generation immigrants								trends							
	1995		1999		2003		2007		95->99		99->03		03->07		95->07	
	Perc.	SE	Perc.	SE	Perc.	SE	Perc.	SE	%	sign	%	sign	%	sign	%	sign
Armenia					8	0.5	7	0.5					−1	⇧		
Australia	28	0.9	28	1.0	31	1.6	29	0.8	0	⇧	3	⇧	−2	⇧	2	⇧
Austria	9	0.5														
Bahrain					14	0.5	12	0.5					−1	⇧		
Belgium (Flemish)	10	0.8	8	0.8	12	1.0			−1	⇧	4	⇧				
Belgium (French)	27	1.2														
Bosnia and Herzegovina							5	0.5								
Botswana					7	0.4	10	0.6					4	⇩		
Bulgaria			3	0.4	2	0.2	2	0.2			−1	⇧	0	⇧		
Canada			20	0.7												
Canada (British Columbia)							29	1.3								
Canada (Ontario)					29	1.4	30	1.5					1	⇧		
Canada (Quebec)					13	1.1	16	1.6					2	⇧		
Chile			2		2	0.2	2	0.3			0	⇧				
Chinese Taipei			4	0.3	3	0.3	2	0.3			−1	⇨	0	⇧		
Colombia	3	0.4													0	⇧
Cyprus	5	0.3	5	0.4	8	0.4	10	0.4	1	⇧	2	⇦	3	⇦	6	⇦
Czech Republic	8	0.6	7	0.8			7	0.4	0	⇧					−1	⇧
Denmark	7	0.5					7									
Egypt					9	0.6	4	0.3					−6	⇨		
El Salvador							3	0.3								
England	14	1.5	17	1.8	13	1.6	14	1.1	3	⇧	−4	⇧	1	⇧	0	⇧

Country	%	(SE)	%	(SE)	%	(SE)	%	(SE)	Diff		Diff		Diff		Diff	
Estonia																
Finland			1	0.2												
France																
Georgia							4	0.6								
Germany	10	0.7			23	1.0							−1	⇑		
Ghana	5	0.3			6	0.4	5	0.4								
Greece																
Hong Kong, SAR	45	1.0	40	1.1	32	0.9	35	1.0	−6	⇒	−7	⇒	3	⇐	−11	⇒
Hungary	2	0.3	2	0.3	2	0.3	4	0.4	0	⇑	0	⇑	2	⇐	−1	⇐
Iceland	4	0.4														
Indonesia			1	0.3	2	0.2	1	0.2			1	⇐	−1	⇒		
Iran, Islamic Republic of	3	0.4	3	0.4	3	0.4	2	0.3	−1	⇑	1	⇑	−1	⇑	−1	⇑
Ireland	8	0.4														
Israel	45	2.3	34	1.1	27	1.1	26	0.9	−11	⇒	−7	⇒	−1	⇑	−19	⇒
Italy			5	0.4	6	0.4	7	0.5			1	⇑	1	⇑		
Japan					1	0.1	1	0.2					1	⇐		
Jordan			32	1.3	22	1.0	22	1.0			−9	⇒	0	⇑		
Korea, Republic of			0	0.1	0	0.1	0	0.1			0	⇑	0	⇑		
Kuwait	28	2.1			25	1.0	14	0.8			13	⇐			−14	⇒
Latvia	12	0.5	13	0.9					1	⇑						
Lebanon					6	0.5	6	0.5					0	⇑		
Lithuania	9	0.5	9	0.6	9	0.7	6	0.5	1	⇑	0	⇑	−3	⇒	−2	⇒

(continued)

Table 4A.3 (continued)

Country	2nd generation immigrants								trends							
	1995		1999		2003		2007		95->99		99->03		03->07		95->07	
	Perc.	SE	Perc.	SE	Perc.	SE	Perc.	SE	%	sign	%	sign	%	sign	%	sign
Macedonia			9	0.9	8	0.8					-2	⇧				
Malaysia			4	0.3	5	0.5	5	0.4			0	⇧	0	⇧		
Malta							12	0.5								
Moldova			12	1.1	12	0.9		0.8			0	⇧				
Mongolia							11									
Morocco			3	0.3	6	0.7	6	0.5			3	⇦	0	⇧		
Netherlands	11	1.2	13	1.6	13	1.1			2	⇧	1	⇧				
New Zealand	20	0.8	20	1.0	23	1.2			0	⇧	2	⇧				
Norway	6	0.4			9	0.6	11	0.7					2	⇧	4	⇦
Oman							9	0.6								
Palestinian National Authority					6	0.4	7	0.5					1	⇧		
Philippines	3	0.3	3	0.3	2	0.3			0	⇧	0	⇧				
Portugal	7	0.6														
Qatar							25	0.5								
Romania	15	1.1	6	2.1	1	0.1	1	0.2	-10	⇨	-5	⇨	0	⇧	-15	⇨
Russian Federation	12	1.0	11	0.7	10	0.6	10	0.7	-1	⇧	-2	⇧	0	⇧	-2	⇧
Saudi Arabia					12	1.0	12	0.8					0	⇧		
Scotland	14	0.7			5	0.4	7	0.5					2	⇦	-7	⇨
Serbia					12	0.6	16	0.8					4	⇦		
Singapore	25	0.6	23		19	0.5	19	0.6	-2	⇨	-4	⇨	1	⇧	-6	⇨
Slovak Republic	7	0.4	7	0.6	7	0.5			0	⇧	0	⇧				
Slovenia	17	0.7	18	1.2	17	1.1	15	1.0	1	⇧	-1	⇧	-1	⇧	-2	⇧

Country																
South Africa	7	0.4							-3	⇨	1	⇦				
Spain	12	0.7	4	0.3	6	0.5	5	0.5					2	⇦		
Spain (Basque Country)					4	0.4										
Sweden	13	0.8			16	1.1	17	1.0					2	⇧	5	⇦
Switzerland	22	0.8					6	0.4								
Syria, Arab Republic of					0	0.1							6	⇦		
Thailand	1	0.2	1	0.2	1	0.2	2	0.3	0	⇧					1	⇦
Tunisia			3	0.3			4	0.3			-2	⇨	3	⇦		
Turkey			3	0.3			2	0.3								
Ukraine							17	0.9								
United Arab Emirates (Dubai)							31	1.4								
United States	13	1.1	15	0.9	13	0.8	18	1.1	2	⇧	-2	⇨	5	⇦	5	⇦
United States (Indiana)					5	0.6										
United States (Massachusetts)							17	1.5								
United States (Minnesota)							10	1.2								

The increase of second-generation immigrant students can pose a challenge to the educational system. As previous research has shown, in general, immigrant students tend to perform less well than native students (Martin, Mullis, Foy, Arora, & Stanco 2012; Mullis, Martin, Foy, Arora, & Stanco 2012; OECD, 2010c). I will analyze this in more detail in the following.

Trends in Mathematics Achievement for Immigrant Students

I will now come to analyzing the achievement differences between immigrant students and native students. TIMSS does not only deliver background information that enables us to identify first- and second-generation immigrant students and also some of their background characteristics. TIMSS also delivers reliable achievement measures for mathematics and science. The mathematics and science scales are calculated using the current as well as the previous study cycles which lead to measures that cannot only be compared within one cycle but also across study cycles, as already explained in the method chapter.

Table 4A.4 shows the trends in performance for immigrant students in TIMSS in mathematics. It displays the mean mathematics achievement for native students in each TIMSS cycle for all countries that participated in the cycle. Next to these mean achievements, the difference for first- and second-generation immigrant students is displayed. A negative number indicates that the immigrant students were performing low than native students. Colored fields indicate that the difference is statistically significant.

As can be seen in Table 4A.4, the number of countries where the immigrant students are statistically significantly outperformed in mathematics by native students increases from 1995 to 2007. This is not only an effect of increased country participation but also of increased differences in participating countries. In TIMSS 1995 in 17 out of 37 participating countries, first-generation immigrant students were outperformed statistically significantly by native students, and in 10 out of 37 countries, second-generation immigrant students were outperformed by native students. In TIMSS 1999, the number of countries where first-generation immigrant students were outperformed statistically significantly by native students which decreased to 12 out of 37 countries, and in only 6 out of the 37 participating countries, second-generation immigrant students were outperformed by native students. In TIMSS 2003, the number of countries where first-generation immigrant students were outperformed statistically significantly by native students increased to 39 out of 51 countries, and only in 12 countries there was no statistically significant difference. The number of countries where second-generation immigrant students were outperformed by native students increased to 16 out of 51 countries. In TIMSS 2007 even in 42 out of 55 countries, first-generation immigrant students were outperformed statistically significantly by native students. And in 21 out of 55 countries, second-generation immigrant students were outperformed statistically significantly by native students.

Table 4A.4 Trends in immigrant students' mathematics achievement

	1995				1999				2003				2007			
	Native		Difference of immigrants		Native		Difference of immigrants		Native		Difference of immigrants		Native		Difference of immigrants	
Country	Math Ach	SE	1st gen	2nd gen	Math Ach	SE	1st gen	2nd gen	Math Ach	SE	1st gen	2nd gen	Math Ach	SE	1st gen	2nd gen
Armenia									481	3.1	−12	−11	498	2.9	9	5
Australia	512	3.7	9	3	524	5.3	4	3	502	4.5	16	11	496	3.8	1	4
Austria	530	2.6	−55 ⇨	−22 ⇨				⇨								
Bahrain									405	1.9	−27 ⇨	5	403	1.9	−26 ⇨	3
Belgium (Flemish)	565	4.2	−16	−25 ⇨	562	2.9	−44	−19	549	2.6	−64 ⇨	−51 ⇨				⇨
Belgium (French)	527	3.4	−36 ⇨	−21 ⇨				⇨								
Bosnia and Herzegovina													457	2.9	−4	5
Botswana									369	2.5	34	−9	369	2.3	−22 ⇨	−28 ⇨
Bulgaria					511	6.1	8	−4	481	4.1	−66 ⇨	−3	473	4.8	−65 ⇨	−10 ⇨
Canada					533	2.9	−10	−5								
Canada (British Columbia)													499	2.7	35 ⇦	18 ⇦
Canada (Ontario)									519	2.7	11	3	512	4.5	22 ⇦	9
Canada (Quebec)									546	3.2	−30 ⇨	−7	531	3.2	−17	−1
Chile					394	4.3	−12	−11	390	3.3	−48 ⇨	−24 ⇨				⇨
Chinese Taipei					587	4.0	−42	−16	589	4.4	−109 ⇨	−42 ⇨	606	4.3	−111 ⇨	−15 ⇨
Colombia	376	3.0	16 ⇦	3									385	3.5	−67 ⇨	−41 ⇨
Cyprus	461	1.6	−2	0	478	1.9	−2	−13	465	1.6	−26 ⇨	−1	472	1.7	−44 ⇨	−6 ⇨
Czech Republic	545	3.9	−15	−24 ⇨	522	3.9	−19	−23 ⇨								
Denmark	488	2.1	−31 ⇨	−9								⇨	505	2.5	−19 ⇨	−14 ⇨

(continued)

Table 4A.4 (continued)

Country	1995 Native Math Ach	1995 SE	1995 1st gen	1995 2nd gen	1999 Native Math Ach	1999 SE	1999 1st gen	1999 2nd gen	2003 Native Math Ach	2003 SE	2003 1st gen	2003 2nd gen	2007 Native Math Ach	2007 SE	2007 1st gen	2007 2nd gen
Egypt									423	3.5	−40 ⇨	−34 ⇨	427	3.3	−76 ⇨	−67 ⇨
El Salvador													344	2.9	−47 ⇨	−3 ⇨
England	492	2.4	−19 ⇨	4	497	4.6	−5	4	502	5.4	−23	9	515	5.1	−20	13
Estonia									534	3.4	−29 ⇨	−6				
Finland					522	2.7	−38 ⇨	−21								
France																
Georgia													418	6.0	−47 ⇨	−55 ⇨
Germany	503	4.3	−40 ⇨	−22 ⇨					298	4.9	−59 ⇨	−16 ⇨				
Ghana													324	4.3	−60 ⇨	−24 ⇨
Greece	464	2.7	−14 ⇨	−20 ⇨												
Hong Kong, SAR	570	7.5	12	13	578	4.7	2	12	590	3.9	−15 ⇨	4	579	5.9	−26 ⇨	4
Hungary	520	3.1	−19	8	533	3.7	−27	−9	531	3.2	−22	20	519	3.3	−58 ⇨	4 0.2
Iceland	474	2.6	−1	0												
Indonesia					405	4.7	−60	−64 ⇨	422 ⇨	4.9	−58 ⇨	−50 ⇨	409	3.7	−58 ⇨	−61 ⇨
Iran, Islamic Republic of	415	1.8	−14	−21 ⇨	423	3.5	−16	−4	414	2.3	−43 ⇨	−26 ⇨	405	4.1	−29	−38 ⇨
Ireland	514	3.4	8	−4												
Israel	532	8.3	−16	−14	462	3.3	5	16 ⇦	500 ⇦	4.0	−18 ⇨	5	469	4.1	−27 ⇨	11
Italy					481	3.9	−25 ⇨	−6	486	3.2	−29 ⇨	−11	481	3.2	−30 ⇨	−1
Japan					579	1.7	−37 ⇨		571	2.1	−35 ⇨	−25 ⇨	571	2.4	−34	−10
Jordan					412	4.0	34 ⇦	35 ⇦	433 ⇦	3.8	−36 ⇨	13	430	4.7	−45 ⇨	21
Korea, Republic of					587	2.0	−4	−46 ⇨	591 ⇨	2.2	−91 ⇨	−63 ⇨	598	2.7	36	−66 ⇧

	1	2	3	4	5	6	7	8	9	10	11	12	13	14	15	16	17	18	19	20	21	22	23
Kuwait	387	2.7	22	⇦	12	⇦	506	3.7	−9		−6	509	3.3	−16		6		362	2.7	−33	⇨	2	⇨
Latvia	478	2.5	−23	⇨	−7							436	3.1	−6		−7		459	4.0	−36	⇨	−6	⇨
Lebanon																		510	2.3	−64	⇨	−1	⇨
Lithuania	454	2.8	4		0		480	4.4	−25			504	2.7	−48	⇨	4							
Macedonia							450	4.2	−24	⇦	16	446	3.6	−63	⇨	−19							⇨
Malaysia							520	4.4	−1		−15	511	4.2	−21	⇨	−24	⇨	479	5.0	−50	⇨	−27	⇨
Malta											−12							494	1.5	−62	⇨	−2	⇨
Moldova							467	4.0	15		13	463	4.1	−17	⇨	8							
Mongolia																		448	3.8	−58	⇨	−52	⇨
Morocco							340	3.2	−30	⇨	−12	395	2.5	−38	⇨	−27	⇨	388	2.9	−65	⇨	−28	⇨
Netherlands	533	5.7	−37	⇨	−15		546	6.9	−25	⇨	−24	544	3.9	−39	⇨	−32	⇨						
New Zealand	489	3.0	7		1		488	5.0	19	⇦	4	493	5.2	27	⇦	3	⇨						
Norway	484	2.0	−28	⇨	−3							467	2.4	−40	⇨	−9							
Oman																		474	2.2	−29	⇨	−12	⇨
Palestinian National Authority												401	3.1	−32	⇨	−16	⇨	387	3.4	−70	⇨	−20	⇨
Philippines	394	2.2	−26	⇨	−12	⇨	362	6.0	−72	⇨	−60	383	5.2	−71	⇨	−51	⇨	380	3.5	−51	⇨	−12	⇨
Portugal	438	2.2	−2		6																		
Qatar																		305	2.1	−9	⇨	26	⇦
Romania	469	3.8	4		−1		477	5.9	−61	⇨	−24	478	4.8	−50	⇨	−50	⇨	466	4.1	−85	⇨	−94	⇨
Russian Federation	518	4.0	−6		8		529	5.9	−20		−11	511	4.0	−14	⇨	4		515	3.9	−15		−12	
Saudi Arabia												333	4.9	−14		11		335	3.0	−37	⇨	16	⇦
Scotland	477	3.6	19	⇦	21	⇦						501	3.7	−36	⇨	10		491	3.6	−41	⇨	8	⇨
Serbia												483	2.5	−36	⇨	0		488	3.5	−31	⇦	8	⇨
Singapore	620	5.1	11		6		600	6.2	23	⇦	9	607	3.6	−18	⇨	7		588	3.9	34	⇦	9	⇨

(continued)

Table 4A.4 (continued)

Country	1995 Native Math Ach	SE	Difference of immigrants 1st gen	2nd gen	1999 Native Math Ach	SE	Difference of immigrants 1st gen	2nd gen	2003 Native Math Ach	SE	Difference of immigrants 1st gen	2nd gen	2007 Native Math Ach	SE	Difference of immigrants 1st gen	2nd gen
Slovak Republic	527	2.8	4	−4	535	3.7	−45 ⇨	−2	512	3.0	−64 ⇨	−18				
Slovenia	520	2.7	−24 ⇨	−4	534	3.1	−27 ⇨	−15 ⇨	498	2.4	−19 ⇨	−21 ⇨	509	2.3	−61 ⇨	−21 ⇨
South Africa	353	3.5	−19 ⇨	0	283	7.2	−59 ⇨	−9	292	7.3	−78 ⇨	−19				
Spain	470	2.0	−17 ⇨	−8												
Spain (Basque Country)									491	2.7	−48 ⇨	−12	505	2.8	−57 ⇨	−13
Sweden	523	2.3	−44 ⇨	−13 ⇨					507	2.6	−48 ⇨	−18	499	2.2	−41	−16 ⇨
Switzerland	557	2.1	−56 ⇨	−22 ⇨												
Syria, Arab Republic of									334	17.8	21	−11	409	3.4	−45 ⇨	−23 ⇨
Thailand	509	5.0	−38 ⇨	4	469	5.2	−33 ⇨	9					443	4.9	−62 ⇨	−28 ⇨
Tunisia					448	2.4	2	−4	411	2.2	−16	13	423	2.5	−17 ⇨	−29 ⇨
Turkey					429	4.2	−1	−1					434	4.8	−29	−30
Ukraine													466	3.6	−69 ⇨	12 ⇦
United Arab Emirates (Dubai)													397	5.7	91 ⇦	65 ⇦
United States	490	4.1	−18 ⇨	−10	510	3.7	−44 ⇨	−18	512	3.0	−53 ⇨	−17 ⇨	517	2.8	−49 ⇨	−19 ⇨
United States (Indiana)									510	5.0	−35 ⇨	14 ⇨				
United States (Massachusetts)													558	4.0	−61 ⇨	−25 ⇨
United States (Minnesota)													539	4.1	−53 ⇨	−24 ⇨

In terms of trends, this means that for first-generation immigrant students, the situation worsened: in TIMSS 1995, they achieved statistically significantly lower in 46 % of the participating countries, but this figure rose – with a slight improvement in 1999 – to about 76 % of participating countries in TIMSS 2003 and 2007. Also for second-generation immigrant students, the situation improved from being statistically significantly outperformed in terms of their achievement in 27 % of the participating countries in 1995 to only 16 % in 1999, but it worsened to achieving statistically significantly lower in 31 % in 2003 and even 41 % of the participating countries in 2007.

Another perspective that could be taken is to compare the average achievement differences for immigrant and native students instead of the pure number of countries with statistically significant differences. My comparison reveals that across all 37 participating countries in TIMSS 1995, in mathematics, first-generation immigrant students were outperformed by native students by 14 score points. This average increased to 19 score points in 1999, to 33 in 2003, and to 35 in 2007. For second-generation immigrant students, I find an average difference across all countries of 5 score points in TIMSS 1995, 9 in 1999, 11 in 2003, and 12 in 2007. This means that for both groups of immigrant students, the average difference across all countries increases between the different TIMSS cycles.

Comparing the achievement differences, one could also focus only on the countries where there is a statistically significant difference between immigrant students and native students. In TIMSS 1995, I observe an average difference between native students and first-generation immigrant students among the 17 countries that have shown statistically significant differences of 31 score points for these two groups. In TIMSS 1999, this average difference for the 12 countries that show a statistically significant difference increased to 41 score points. In 2003 the difference for the 39 countries with statistically significant differences increased to 43 score points and in TIMSS 2007 even to 49 score points for the 42 countries with statistically significant differences.

For second-generation immigrant students, the trends are similar. In TIMSS 1995, I determine an average difference between native students and second-generation immigrant students among the ten countries that have statistically significant differences of 20 score points between native and second-generation immigrant students. In TIMSS 1999, this average difference for the six countries that show a statistically significant difference increased to 38 score points. In 2003 the difference for the 16 countries with statistically significant differences then decreased to 31 score points, but in TIMSS 2007, it increased to 34 score points for the 21 countries with statistically significant differences.

I conclude that examining the percentage of countries that show statistically significantly lower achievement for first- and second-generation immigrant students compared to native students, an increase of these among TIMSS participants is detected. Also when looking at the actual average differences, these increase for first-generation immigrant students as well as for second-generation immigrant students in the TIMSS cycles – independently of how I calculated these differences.

Now, I want to look more in-depth into the results and investigate the results for individual countries.

In Belgium (Flemish), there is a steady decline of the mathematics achievement of first-generation immigrant students. In 1995, they achieved 16 score points below native students' score points, in 1999 44 score points, and in 2003 64 score points. Together with the increase of the percentage of first-generation immigrants between 1995 and 2003 from 3 % to 4 % to 7 %, this indicates a clear challenge for the educational system. The situation for second-generation immigrant students is somewhat similar. They were lagging behind native students by 25 score points in 1995, by 19 score points in 1999, and by 51 score points in 2003. Other countries with a clear increase of the achievement gap of first-generation immigrant students are Chile, Chinese Taipei, Colombia, Cyprus, Egypt, Hong Kong, Hungary, Jordan, Lebanon, Lithuania, Macedonia, Malaysia, Morocco, the Philippines, Romania, Saudi Arabia, Scotland, Slovak Republic, Slovenia, South Africa, Spain, Syria, Thailand, Tunisia, Turkey, and the United States.

For some other countries, the results for the different years are changing quite substantially. The reason for this is probably that the group of immigrants assessed was very small. An example is Korea where first-generation immigrant students were lagging behind by 4 points in 1999, by 91 points in 2003 but outperformed native students by 36 points in 2007. But as seen in Table 4A.1, the percentage of first-generation immigrant students in Korea was 1.9 % in 1999, 1.4 % in 2003, and 0.5 % in 2007. The results for countries with such a small percentage of immigrants in the population and consequently in the sample should be interpreted very carefully because results from very few observations can lead to large differences and should not be over-interpreted.

There are also positive examples of countries where the achievement of immigrant students was more equal to native students or at least showed a positive trend. In Australia both immigrant student groups were performing at about the same level as native students, i.e., even slightly better. Considering that there are between 10 and 15 % of first-generation immigrant students and about 30 % of second-generation immigrant students, in terms of delivering equal opportunities, Australia seems to be quite successful. Further research on the background and the potential reasons for this will be conducted below.

But also in the Canadian province of Ontario, both groups of immigrant students were achieving statistically significantly better than native students in the two TIMSS cycles in which they participated. The same is true for the Canadian province of British Columbia in TIMSS 2007, but interestingly, the opposite is the case for Quebec.

In Hong Kong the achievement of first and second generation immigrants was similar to the achievement of native students in 1995 and 1999. In 2003 and 2007, the achievement of first-generation immigrant students became statistically significantly lower than that of the native students. At the same time, the percentage of immigrant students decreased. Overall there is a slight achievement increase for native students between 1995 and 2007, while on the other hand, I observe a decrease of about 30 score points for first-generation immigrant students.

In New Zealand, second-generation immigrant students achieved the same level as native students. First-generation immigrant students even achieved 7 points above native students' score points in 1995, 19 points in 1999, and 27 points in 2003 – the latter means a statistically significant difference favoring the immigrant students.

Also Singapore has a case where immigrant students achieved positive performance with the only exception in the 2003 cycle in which first-generation immigrant students were outperformed statistically significantly by native students. In all other cycles, immigrant students – of the first as well as second generation – achieved the same level as native students or even outperformed them.

It becomes apparent that overall, immigrant students are performing less well in mathematics than native students. This is particularly true for first-generation immigrant students. In terms of the trends, I also observe that there is an increasing number of countries with a statistically significantly lower performance of immigrant students in mathematics. Moreover the magnitude of the differences increases over time. However, I find some countries with rather positive results – also with regard to achievement trends, i.e., the Canadian provinces of British Columbia and Ontario, Singapore, Dubai, and to a certain extent Australia.

Trends in Science Achievement for Immigrant Students

TIMSS assesses not only the mathematics but also science performance of students. Following the analysis of trends in mathematics achievement, I will now turn to investigate the science achievement. TIMSS provides scores on student levels not only for mathematics but also for science, and the scaling approach enables me to compare differences not only within one study cycle but also across study cycles. Table 4A.5 shows the trends in science achievements for native students and first- and second-generation immigrant students.

The same tendency that is observed for mathematics achievement also applies to science achievement. In 17 out of 37 participating countries in the 1995 assessment, native students outperformed first-generation immigrant students statistically significantly. In 12 countries, native students outperformed second-generation immigrant students statistically significantly. In 1999 the situation is very similar: in 16 out of 38 participating countries, native students outperformed first-generation immigrant students statistically significantly, and in nine countries, second-generation immigrant students were outperformed statistically significantly. In 2003 the statistics change negatively for immigrant students, with first-generation immigrant students being outperformed statistically significantly in 41 out of 51 countries, which also holds true for second-generation immigrant students in 20 countries. Quite similarly in 2007, in 44 out of 56 countries, first-generation immigrant students and, in 23 countries, second-generation immigrant students were out-performed statistically significantly.

Percentagewise this means an increase in the number of countries where statistically significant disadvantages for first-generation immigrants among native students could be observed from about 46 % in 1995 to 79 % in 2007. For second-generation immigrants, the trend is less negative and the number of countries with statistically significantly lower achievement of second-generation immigrant students increased only from 32 to 41 %.

As for the trends in absolute numbers, I find that the negative performance difference in the average number of score points of first-generation immigrant students across all countries increased from 18 score points in 1995 to 23 in 1999 to 35 in 2003 up to 39 in 2007. For second-generation immigrant students, the trend shows an increase of the achievement difference from 8–10 to 13–14 score points.

The trends in the maximum number of score points that immigrant students are behind native students are alarming. In 1995 Austria showed the maximum difference at 73 score points, followed by Switzerland (66 score points) and Sweden (62 score points). In 2007 the maximum difference increased to 95 score points in Chinese Taipei, followed by 82 score points in Malaysia and 80 in Egypt. The trend for the maximum differences for second-generation immigrant students is more pronounced: in 1995 the maximum difference was observed for Germany with 34 score points followed by Iran with 32 and Austria with 31 score points. In 2007 the maximum difference was 94 for Romania followed by Georgia and Egypt with a difference of 65 score points each.

There are only few cases where immigrant students performed statistically significant better than native students. In 1995 these were Scotland and Kuwait for both immigrant groups and Ireland for first-generation immigrant students. In 1999 these were Moldova and Singapore for first-generation immigrants, Jordan for both immigrant groups, and Israel for second-generation immigrant students. In 2003 it was only Jordan and Scotland where second-generation immigrant students performed statistically better than native students, but none of the first-generation immigrant groups did. In 2007 in Singapore and Korea, first-generation immigrant students performed statistically significantly higher than native students, in Dubai both immigrant groups performed above the native students' performance, and in Qatar and Jordan, second-generation immigrant students outperformed native students statistically significantly.

Answering the final aspects of research question two, I conclude that the situation for science is similar to what I discovered for mathematics. Overall immigrant students are performing less well than native students in science. Again, this is especially true for first-generation immigrant students. The trends show an increasing number of countries with a lower performance of immigrant students in science, and the magnitude of the differences increases over time. The countries with rather positive results, on the other hand, are again Singapore and Dubai, whereas Australia and the Canadian provinces show less positive statistics in science than in mathematics. Jordan, Scotland, and Qatar show partially good results for immigrant students.

Table 4A.5 Trends in immigrant students' science achievement

Country	1995 Native Science Ach	1995 SE	1995 Diff. immigrants 1st gen	1995 Diff. immigrants 2nd gen	1999 Native Science Ach	1999 SE	1999 Diff. immigrants 1st gen	1999 Diff. immigrants SE	1999 Diff. immigrants 2nd gen	2003 Native Science Ach	2003 SE	2003 Diff. immigrants 1st gen	2003 Diff. immigrants SE	2003 Diff. immigrants 2nd gen	2007 Native Science Ach	2007 SE	2007 Diff. immigrants 1st gen	2007 Diff. immigrants SE	2007 Diff. immigrants 2nd gen
Armenia										465	3.5	−18 ⇨	8.5	−10	487	5.3	17	18.9	16
Australia	526	3.4	−9	0	544	4.7	−15	9.6	−6	528	4.1	−6	8.1	6	517	3.5	−10	8.4	−3
Austria	545	2.6	−73 ⇨	−31 ⇨															⇨
Bahrain										441	2.2	−26 ⇨	4.9	5	474	1.8	−40 ⇨	5.5	7
Belgium (Flemish)	541	2.8	−8	−22 ⇨	539	2.6	−31	16.9	−25 ⇨	527	2.3	−55 ⇨	10.0	−51 ⇨					
Belgium (French)	466	2.9	−23 ⇨	−19 ⇨															
Bosnia and Herzegovina															467	2.9	−1	5.4	−3
Botswana					519	5.5	6	17.8	−11	367	2.7	46	24.6	−9	362	3.1	−46 ⇨	12.4	−39 ⇨
Bulgaria										482	5.2	−41 ⇨	12.6	16	480	5.6	−56 ⇨	13.2	−24
Canada					540	2.7	−36	7.2	−13 ⇨	⇨									
Canada (British Columbia)															526	2.6	1	7.2	1
Canada (Ontario)										538	2.4	−15 ⇨	7.1	−8	526	4.8	1	7.7	3
Canada (Quebec)										538	2.9	−55 ⇨	8.1	−22 ⇨	512	2.9	−21	8.3 ⇨	−13
Chile					422	3.7	−30	12.8	−16	416	2.9	−51 ⇨	6.6	−28 ⇨					
Chinese Taipei					570	4.2	−46	21.8	−10	574	3.3	−90 ⇨	13.0	−26 ⇨	568	3.4	−95	10.1 ⇨	−12
Colombia	399	3.6	2	−17											422	3.5	−64	10.1 ⇨	−40 ⇨

(continued)

Table 4A.5 (continued)

Country	1995 Native Science Ach	1995 SE	1995 1st gen	1995 2nd gen	1999 Native Science Ach	1999 SE	1999 1st gen	1999 SE	1999 2nd gen	2003 Native Science Ach	2003 SE	2003 1st gen	2003 SE	2003 2nd gen	2007 Native Science Ach	2007 SE	2007 1st gen	2007 SE	2007 2nd gen
Cyprus	442	1.6	1	9	460	2.3	9	8.2	5 ⇨	445	2.1	−11	6.3	−1	456	2.1	−30	7.2	−5 ⇨
Czech Republic	555	2.9	−9	−21 ⇨	541	4.4	−25	16.2	−22 ⇨						541	2.0	−31	8.4	−11 ⇨
Denmark	464	2.4	−36 ⇦	−10															
Egypt										442	3.8	−48	7.3	−42 ⇨	446	3.7	−80	6.1	−65 ⇨
El Salvador															391	3.0	−46	8.4	−5 ⇨
England	536	2.9	−23	−3	542	5.0	−12	14.9	−11	549	4.5	−33	11.3	−2	546	4.5	−32	11.8	2
Estonia										559	2.6	−45	8.5	−17 ⇨					
Finland					538	3.3	−70	20.5	−29 ⇨										
France																			
Georgia	526	4.1	−59 ⇨	−34 ⇨											431	4.6	−60	14.5	−65 ⇨
Germany	474	2.2	−12	−15 ⇨															
Ghana										287	5.9	−82	9.5	−46 ⇨	321	5.1	−73	9.1	−36 ⇨
Greece																			
Hong Kong, SAR	504	5.1	10	9	527	4.5	5	7.8	7	556	3.3	−8	6.2	8	533	5.4	−15	8.7	5
Hungary	535	2.7	−27 ⇨	8	553	3.7	−8	17.3	−15	545	2.9	−23	9.8	16	541	2.7	−56	17.1	2
Iceland	478	2.5	−2	−2															
Indonesia					437	4.4	−61	31.0	−51 ⇨	429	4.0	−45	7.4	−47 ⇨	437	3.2	−55	6.6	−50 ⇨
Iran, Islamic Republic of	454	2.3	−24	−32 ⇨	449	3.7	−25	23.8	−18	455	2.3	−47	11.6	−23 ⇨	460	3.6	−16	18.5	−41 ⇨
Ireland	515	3.0	20 ⇦	0															
Israel	542	8.1	−29	−20 ⇨	464	4.7	4	15.1	18 ⇨	493	3.5	−16	7.5	4 ⇦	476	4.5	−32	9.0	6 ⇨
Italy					494	3.8	−29	13.9	−5 ⇨	494	3.1	−35	7.6	−13	497	3.0	−33	7.5	1 ⇨
Japan					550	2.2	−26	19.8		553	1.7	−38	12.5	−17	555	1.9	−45	19.0	−9 ⇨

Jordan	423					436	4.8	33	⇦	7.6	31	⇦	484	4.0	-41	⇨	7.0	15	⇦	487	4.2	-54	⇨	9.7	16	⇦
Korea, Republic of	461					549	2.6	-9	⇦	13.9	-11	⇦	560	1.6	-77	⇨	13.7	-54	⇦	553	2.0	52	⇨	23.8	-27	
Kuwait		4.6	19	⇦	17	503	4.8	-4		14.0	-4	⇦	513	2.6	-9		8.3	3		428	2.9	-45	⇨	7.3	-2	⇨
Latvia		2.2	-29	⇨	-4								397	4.5	-4		11.1	-8								
Lebanon																				425	6.1	-41	⇨	9.1	0	
Lithuania	442	3.0	-5		-5	488	4.3	-39	⇨	16.4	12		521	2.3	-36	⇨	8.3	2		522	2.5	-64	⇨	9.6	-3	
Macedonia						460	5.6	-36	⇨	18.0	-9		461	3.5	-62	⇨	9.3	-22	⇨							
Malaysia						492	4.5	11		19.1	-2		513	3.7	-21	⇨	7.3	-20	⇨	479	5.8	-82	⇨	12.6	-25	⇨
Malta																				462	1.5	-54	⇨	8.2	-4	
Moldova						457	4.4	29	⇦	8.5	3	⇦	475	3.4	-15	⇨	6.5	12								
Mongolia																				463	3.0	-54	⇨	6.2	-42	⇨
Morocco						327	3.9	-33	⇨	12.6	-37	⇨	404	2.6	-37	⇨	6.4	-13		408	3.0	-51	⇨	14.0	-33	⇨
Netherlands	544	4.7	-45	⇨	-13	555	6.2	-47	⇨	16.2	-43	⇨	544	3.2	-41	⇨	6.9	-34	⇨							
New Zealand	505	3.2	-6		1	511	4.9	-11		10.1	4		521	4.8	10		9.5	2								
Norway	508	1.7	-41	⇨	-11								501	1.8	-56	⇨	8.0	-16	⇨	493	2.4	-45	⇨	6.8	-23	⇨
Oman																				437	2.9	-70	⇨	7.2	-17	⇨
Palestinian National Authority													448	3.0	-37	⇨	6.2	-21	⇨	420	3.5	-62	⇨	8.1	-24	⇨
Philippines	391	3.4	-40	⇨	-26	366	7.8	-89	⇨	11.2	-74	⇨	384	5.8	-88	⇨	16.6	-60	⇨							
Portugal	453	2.2	-2		5																					
Qatar																				310	3.0	3		4.2	40	⇨
Romania	472	4.6	-5		-8	474	5.8	-32		30.1	6		472	4.9	-26	⇨	14.9	-25		466	3.8	-68	⇨	11.8	-94	⇨
Russian Federation	509	3.8	-1		10	532	6.1	-15		12.9	-10		517	3.9	-17	⇨	6.3	3		532	3.8	-16	⇨	9.7	-9	
Saudi Arabia													402	4.0	-38	⇨	9.3	6	⇨	411	2.5	-44	⇨	7.0	10	⇨
Scotland	487	3.6	30	⇦	27								514	3.3	-36	⇨	9.5	15	⇦	499	3.2	-44	⇨	9.3	11	⇨

(continued)

Table 4A.5 (continued)

Country	1995 Native Science Ach	1995 SE	1995 Diff. 1st gen	1995 Diff. 2nd gen	1999 Native Science Ach	1999 SE	1999 Diff. 1st gen	1999 Diff. SE	1999 Diff. 2nd gen	2003 Native Science Ach	2003 SE	2003 Diff. 1st gen	2003 Diff. SE	2003 Diff. 2nd gen	2007 Native Science Ach	2007 SE	2007 Diff. 1st gen	2007 Diff. SE	2007 Diff. 2nd gen
Serbia										471	2.7	−18 ⇨	5.9	5	473	3.4	−32 ⇨	9.2	7
Singapore	573	5.5	8	10	563	8.1	24 ⇦	11.7	11	580	4.2	−25 ⇨	9.4	9	563	4.7	19 ⇦	7.7	12
Slovak Republic	527	2.8	1	0	536	3.2	−41 ⇨	12.2	3 ⇨	520	3.0	−51 ⇨	8.7	−18 ⇨					
Slovenia	547	2.3	−25 ⇨	−15 ⇨	539	3.5	−33 ⇨	11.3	−22 ⇨	526	2.1	−18 ⇨	6.9	−25 ⇨	545	2.2	−61 ⇨	7.9	−23 ⇨
South Africa	328	5.1	−32 ⇨	−6	253	8.7	−75 ⇨	10.4	−3	279	8.6	−96 ⇨	10.2	−20					
Spain	498	2.0	−8	−4															
Spain (Basque Country)										492	2.8	−43 ⇨	11.5	−16 ⇨	503	3.1	−50 ⇨	9.3	−7
Sweden	539	2.3	−62 ⇨	−14 ⇨						535	2.5	−62 ⇨	7.2	−27 ⇨	521	2.4	−56 ⇨	6.9	−27 ⇨
Switzerland	534	2.0	−66 ⇨	−23 ⇨											467	2.5	−49 ⇨	5.2	−22 ⇨
Syria, Arab Republic of										388	17.1	21	17.5	−10					
Thailand	510	3.2	−25 ⇨	−6	483	4.1	−27 ⇨	11.4	16						472	4.3	−58 ⇨	25.6	−32 ⇨
Tunisia					430	3.5	−2	12.1	1	405	2.1	−17	15.8	−9	447	2.2	−24 ⇨	7.1	−27 ⇨
Turkey					433	4.4	−5	13.7	−9						456	3.8	−28 ⇨	14.8	−29
Ukraine															490	3.4	−70 ⇨	8.2	7
United Arab Emirates (Dubai)															433	5.7	78 ⇦	6.9	60 ⇦
United States	528	4.2	−37 ⇨	−25 ⇨	527	4.1	−68 ⇨	9.8	−30 ⇨	537	2.8	−63 ⇨	6.3	−27 ⇨	532	2.8	−60 ⇨	5.9	−33 ⇨
United States (Indiana)										534	4.4	−47 ⇨	13.0	−3					
United States (Massachusetts)															569	4.2	−74 ⇨	10.2	−33 ⇨
United States															548	4.1	−71 ⇨	14.4	−43 ⇨

Summary

I found in this chapter that the percentage of immigrant students increased in several of the countries analyzed. The data for grade eight appears to be very reliable and further analyses can be performed. On the other hand, the grade four data seems not to be reliable so that further investigation would not seem sensible. The achievement of immigrant students in mathematics and science lags behind the achievement of native students in most countries. The overall trend is that immigrant students are increasingly lagging behind and the results for mathematics and science are very similar in terms of these outcomes. Consequently, for future analysis, I will focus on mathematics only.

I noted that the percentage of first-generation immigrant students in grade eight increased in several countries between 1995 and 2007 – especially between 1999 and 2003, I identify a substantial number of countries with increases in the first-generation immigrant population. For grade eight second-generation immigrant students, I cannot observe such a trend. For grade four students, the data does not seem to be sufficiently reliable to investigate this question.

I found that overall immigrant students – and especially first-generation immigrant students – are performing below native students' performance in mathematics and in science. This applies in particular to first-generation immigrant students. I also found that the trends in immigrant students compared to native students are quite discouraging. In general, immigrant students seem to increasingly lag behind compared to native students' achievement. However, there are some countries where I find positive results and trends in immigrant students, for example, Canada, Singapore, and Dubai.

Chapter 4B
Immigrant Students' Background in TIMSS 2007

Abstract In this chapter attention is paid to the background differences of immigrant students compared to native students in the countries that participated in TIMSS 2007. The students are grouped as native students, first-generation immigrant students, and second-generation immigrant students. Furthermore the achievement levels in mathematics are analyzed for some subgroups of immigrant students and are contrasted to the levels achieved by native students. The variables explored include students' age, showing that in more than half of the study participants, first-generation immigrant students are statistically significantly older than their native peers, the diverse age at immigration and its effect on mathematics achievement, and gender differences that reveal an underrepresentation of girls in the educational systems. Presenting student questionnaires, the chapter further explores the influence of socioeconomic background on achievement and students' attitudes towards school and mathematics as well as their self-esteem, which leads to some interesting results where high self-esteem does not match the students' achievement scores.

Keywords Immigrant's age • Immigrant girls • Socio economic status • SES • Immigrant's attitudes

In this chapter attention shall be paid to the background differences of immigrant students compared to native students in the countries that participated in TIMSS 2007. The students are grouped as native students, first-generation immigrant students, and second-generation immigrant students. Furthermore the achievement levels in mathematics are analyzed for some subgroups of immigrant students and are contrasted to native students.

Students' Age

As discussed in Chap. 3 the TIMSS samples consist of grade eight students in all countries, one important characteristic of the students being their age. In general the mean age of the students varies between countries because of different school starting age and different policies on the students' promotion and retention.

© Springer International Publishing Switzerland 2016   75
D. Hastedt, *Mathematics Achievement of Immigrant Students*,
DOI 10.1007/978-3-319-29311-0_5

Consequently the comparison of the age of immigrant students to the age of native students in TIMSS is an important topic to consider. But due to the differences in policies regarding school entry age and grade repetition, it makes sense to compare the mean age of immigrant students and native students only within countries. A higher mean age of immigrant students compared to the mean age of native students in a country might have different reasons: the immigrant students could have started school in a country where the school entry age was higher; or the student was enrolled in a lower grade when migrating into the country, for example, due to language problems or administrative problems. Another reason could be grade repetition.

In TIMSS 2007 test administrators tracked the students' answers about the month and year of birth and also the month and year of the test. With this information I calculated the student's age at the time of the test. Table 4B.1 shows the mean age for each of the immigrant and native student groups for each country.

Table 4B.1 reveals the mean age of first-generation immigrant students, native students, and second-generation immigrant students for each country. It also displays two indicators that show whether the mean age of first- and second-generation immigrant students is statistically significantly different from the mean age of native students.

In summary, I can see that in 29 out of 56 countries, first-generation immigrant students are statistically significantly older than their native peers in TIMSS 2007 grade eight. Only in Dubai, first-generation immigrant students are statistically significantly younger than their native peers. In 12 out of 56 countries, second-generation immigrant students are statistically significantly older than their native peers. In five countries second-generation immigrant students are statistically significantly younger than their native peers.

Due to the large sample sizes, also small differences can appear as statistically significantly different. For example, the mean age of 14.84 years of native students in Sweden appears to be statistically significantly different from the mean age of second-generation immigrants of 14.77. However this difference is probably not meaningful.

As described in more detail in the method chapter, in TIMSS 1995 two adjacent grades, grade seven and eight, were tested and scaled on the same metric. The difference in mathematics achievement between the two grades varied between seven points in South Africa and 49 points in Lithuania but was on average 32 points (Beaton, Mullis, et al., 1996, Table 1.3).

I am going to use this information to identify thresholds for meaningful differences in the mean age. The assessment experts in TIMSS defined a difference of five points as relevant and designed the study based on this assumption (Olson et al., 2008, Chapter 5). Following Cliffordson and Gustafsson (2010) who state that half of the achievement gain stems from students aging, one could conclude that an age difference of 0.3 years and more should be considered as significant (32 points/2/5 points).

Going by this definition, I conclude that in Bahrain first-generation immigrant students are considerably older than native students. The same is true for Colombia,

Table 4B.1 Age of students by immigration status

Country	First generation immigrant			Native		Second generation immigrant		
	Mean	SE	sig	Mean	SE	Mean	SE	sig
Armenia	14.9	0.03		14.9	0.01	14.8	0.04	
Australia	13.9	0.03		13.9	0.01	13.9	0.02	⇩
Bahrain	14.3	0.04	⇧	14.0	0.01	14.1	0.04	⇧
Bosnia and Herzegovina	14.7	0.02		14.7	0.01	14.8	0.03	
Botswana	15.0	0.07	⇧	14.8	0.02	14.9	0.07	
Bulgaria	15.0	0.04		14.9	0.01	15.0	0.12	
Canada (British Columbia)	13.9	0.03		13.9	0.01	13.8	0.01	
Canada (Ontario)	13.9	0.01		13.8	0.01	13.8	0.01	⇩
Canada (Quebec)	14.3	0.05	⇧	14.2	0.02	14.2	0.04	
Chinese Taipei	14.2	0.02		14.2	0.01	14.2	0.04	
Colombia	14.8	0.15	⇧	14.4	0.05	14.9	0.15	⇧
Cyprus	14.1	0.03	⇧	13.8	0.01	13.8	0.02	
Czech Republic	14.7	0.07	⇧	14.4	0.01	14.5	0.03	⇧
Egypt	14.1	0.03		14.0	0.06	14.0	0.06	
El Salvador	15.3	0.12	⇧	15.0	0.03	15.1	0.07	⇧
England	14.3	0.02		14.2	0.01	14.2	0.01	
Georgia	14.3	0.05	⇧	14.2	0.02	14.2	0.06	
Ghana	16.0	0.08	⇧	15.8	0.05	15.7	0.14	
Hong Kong, SAR	15.0	0.05	⇧	14.1	0.01	14.2	0.02	⇧
Hungary	14.9	0.10	⇧	14.6	0.01	14.5	0.04	
Indonesia	14.4	0.06	⇧	14.2	0.02	14.6	0.18	⇧
Iran, Islamic Republic of	14.5	0.21		14.2	0.02	14.6	0.09	⇧
Israel	14.2	0.03	⇧	14.0	0.01	14.0	0.02	⇧
Italy	14.3	0.05	⇧	13.9	0.01	13.9	0.03	
Japan	14.4	0.04		14.5	0.00	14.5	0.04	
Jordan	14.0	0.03	⇧	13.9	0.01	13.9	0.02	
Korea, Republic of	14.5	0.16		14.3	0.01	14.3	0.13	
Kuwait	14.7	0.05	⇧	14.3	0.02	14.4	0.03	⇧
Lebanon	14.6	0.07	⇧	14.3	0.03	14.5	0.14	
Lithuania	15.0	0.05		14.9	0.01	14.9	0.04	
Malaysia	14.4	0.04		14.3	0.01	14.3	0.02	
Malta	14.2	0.04	⇧	14.0	0.01	14.0	0.02	
Mongolia	15.0	0.04	⇧	14.8	0.03	15.0	0.04	⇧
Morocco	15.1	0.16		14.8	0.04	14.9	0.10	
Norway	13.9	0.02	⇧	13.8	0.01	13.8	0.01	
Oman	14.5	0.06	⇧	14.2	0.02	14.3	0.05	
Palestinian National Authority	14.0	0.03		14.0	0.02	14.0	0.04	
Qatar	14.1	0.03	⇧	13.9	0.01	13.9	0.02	
Romania	15.1	0.05		15.0	0.01	14.9	0.14	

(continued)

Table 4B.1 (continued)

Country	First generation immigrant			Native			Second generation immigrant		
	Mean	SE	sig	Mean	SE	sig	Mean	SE	sig
Russian Federation	14.6	0.07		14.6	0.03		14.5	0.04	
Saudi Arabia	14.6	0.07	⇧	14.3	0.03		14.3	0.04	
Scotland	13.8	0.04		13.7	0.01		13.7	0.03	
Serbia	14.9	0.04		14.9	0.01		14.9	0.01	
Singapore	15.1	0.05	⇧	14.3	0.01		14.3	0.01	
Slovenia	13.8	0.04		13.8	0.01		13.8	0.02	
Spain (Basque Country)	14.5	0.08	⇧	14.1	0.01		14.2	0.06	⇧
Sweden	14.9	0.04		14.8	0.01		14.8	0.01	⇩
Syria, Arab Republic of	14.1	0.04	⇧	13.9	0.02		14.0	0.05	
Thailand	14.6	0.10	⇧	14.3	0.01		14.5	0.10	⇧
Tunisia	14.5	0.09		14.5	0.03		14.7	0.10	
Turkey	14.6	0.19	⇧	14.0	0.02		14.0	0.06	
Ukraine	14.2	0.04		14.2	0.03		14.1	0.04	
United Arab Emirates (Dubai)	14.1	0.04	⇩	14.4	0.04		14.1	0.04	⇩
United States	14.3	0.02		14.3	0.01		14.2	0.02	⇩
United States (Massachusetts)	14.3	0.04	⇧	14.2	0.02		14.2	0.03	
United States (Minnesota)	14.3	0.05		14.3	0.01		14.3	0.03	

El Salvador, Hong Kong, Hungary, Italy, Kuwait, Singapore, Basque, and Turkey. For second-generation immigrants, the same holds for Colombia, Indonesia, and Iran. An interesting case is Dubai where both immigrant groups are substantially younger than native students.

I conclude that in more than half of the countries, first-generation immigrant students are statistically significantly older than their native peers. Second-generation immigrant students are statistically significantly older than their native peers in more than 20 % of the countries that participated in TIMSS 2007. This result is not very surprising since I know from my analysis presented in Chap. 4A that immigrant students are performing less compared to native students in many countries, and the review in Chap. 2 suggested that in some countries, grade repetition of lower-performing students is applied. One could also hypothesize that the transfer from one educational system into another – in some cases in a different language of instruction – could lead to grade repetition of the students concerned.

Age at Immigration

As stated in Chap. 2, not only is the age of immigrant students compared to native students relevant, but there is also a discussion about the effect of the age of immigration on the immigrants' achievement. Next, I will analyze the age of immigration and its effects on achievement. In the TIMSS assessment questionnaires are

B. If you were not born in <country>, how old were you when you came to <country>?

*Fill in **one** circle only*

Older than 10 years old ---------------- ①

5 to 10 years old ---------------------- ②

Younger than 5 years old --------------- ③

Fig. 4B.1 TIMSS 2007 student questionnaire, variable BS4GBRNC (Source: TIMSS 2007 Assessment. Copyright © 2009 International Association for the Evaluation of Educational Achievement (IEA). Publisher: TIMSS & PIRLS International Study Center, Lynch School of Education, Boston College)

administered asking the students not only if they were born in the country of residence but also, if this was not the case, when they came to the country (Fig. 4B.1).

Table 4B.2 shows the percentages of students checking each of the three categories for each country. Of course this information is only available for first-generation immigrants. The mean mathematics achievement for the three groups of first-generation immigrants is also shown in the table.

There are major differences between the countries in terms of the distribution of the age groups when the students came to the country. In 15 countries more than 50 % of the students came to the country before the age of five. Taking into account the review in Chap. 2 where several references were found indicating that an early arrival in the country is beneficial for later success in school, the immigrants in these 15 countries would have an advantage. On the other hand, I find eight countries where more than 40 % of the students arrived after the age of ten. I conclude that the situation of immigrants with respect to the age of immigration is quite diverse between countries.

Based on the literature review, one would expect that the immigrant students who arrived at a younger age would outperform the immigrant students that arrived later. Interestingly this cannot be clearly confirmed. There are participants that follow the hypothesis, such as Quebec, Chinese Taipei, Cyprus, El Salvador, Ghana, Hong Kong, Israel, Japan, Kuwait, Lebanon, Morocco, Palestine, Qatar, Scotland, Slovenia, Sweden, Syria, Ukraine, and the state of Minnesota – not all of these show a statistically significant pattern. Interesting cases are Ontario and Dubai that exhibit the opposite pattern with students being older at the time of immigration and performing better in mathematics than students arriving at a younger age. Most countries show an amorphous pattern, sometimes with the immigrant students in the middle category performing better, and sometimes with the immigrant students in the middle category performing less well than the immigrant students in the other two categories.

So, I cannot find a clear pattern for the age when students came to the country of residence and their mathematics achievement. I do, however, find a tendency in the sense that in 19 countries the mathematics achievement of the immigrants declined due to the age of immigration – sometimes only slightly and far below any statistical significance. There are only two countries with the opposite pattern, but none of them with statistically significant data.

Table 4B.2 Immigrant students' age at immigration and their mathematics achievement

Country	Older than 10 years				5–10 years old				Younger than 5 years			
	perc.	SE	Math Ach	SE	perc.	SE	Math Ach	SE	perc.	SE	Math Ach	SE
Armenia	38	2.6	508	18.0	26	2.9	552	20.0	37	3.5	475	8.8
Australia	32	2.8	500	11.7	35	3.0	489	10.2	33	2.2	505	10.2
Bahrain	33	2.1	369	6.9	27	1.6	384	8.1	40	2.2	379	6.7
Bosnia and Herzegovina	13	1.5	405	7.8	23	1.9	465	7.0	65	2.5	458	6.7
Botswana	22	3.1	370	25.3	42	3.6	329	13.9	36	3.6	373	15.8
Bulgaria	43	3.4	403	15.8	15	2.5	422	18.7	42	3.4	411	11.6
Canada (British Columbia)	35	2.9	531	8.9	32	2.3	549	12.1	33	2.7	524	8.9
Canada (Ontario)	27	3.8	547	10.1	42	4.3	539	7.9	32	2.9	520	8.8
Canada (Quebec)	26	3.8	493	10.5	30	3.1	519	10.9	44	3.8	525	9.5
Chinese Taipei	27	2.9	486	15.3	20	2.4	490	21.4	54	3.5	500	14.2
Colombia	25	4.7	334	17.3	13	3.0	279	15.5	62	6.5	314	12.2
Cyprus	25	2.3	402	11.7	29	2.4	419	10.7	46	2.6	450	7.4
Czech Republic	22	3.3	488	15.4	29	4.6	508	11.9	49	5.0	475	9.9
Egypt	34	1.4	352	6.5	32	1.8	345	7.6	35	2.3	359	5.5
El Salvador	31	4.6	279	17.4	16	2.8	299	21.7	53	4.8	309	14.3
England	31	3.0	487	15.3	33	2.8	517	11.2	36	3.2	477	15.8
Georgia	28	4.3	377	19.2	17	3.5	351	24.9	55	4.9	384	20.3
Ghana	42	2.4	255	9.2	30	1.9	269	8.8	29	1.9	277	8.5
Hong Kong, SAR	26	2.2	550	14.7	35	1.9	552	8.0	39	1.9	558	7.5
Hungary	22	6.6	433	26.6	26	6.0	508	29.7	53	5.8	454	18.5
Indonesia	40	3.3	361	7.9	20	1.8	335	10.7	40	3.1	354	7.3
Iran, Islamic Republic of	22	9.7	403	57.0	40	10.8	353	30.7	38	10.0	415	39.4
Israel	14	1.7	412	15.9	35	3.2	435	10.3	51	3.1	458	10.9
Italy	25	3.3	446	12.6	35	3.5	457	12.4	39	3.1	450	9.4
Japan	14	5.6	469	39.6	19	7.5	487	46.9	68	8.2	572	24.8
Jordan	35	6.1	405	19.8	22	2.6	372	12.2	44	4.5	375	10.2
Korea, Republic of	13	8.0	537	15.8	26	10.1	658	45.1	61	11.4	644	35.9
Kuwait	26	2.1	305	10.3	15	1.5	323	11.8	59	2.3	336	7.2
Lebanon	49	3.0	408	6.4	21	2.3	416	6.9	30	2.7	457	7.0
Lithuania	48	4.1	448	10.5	15	4.0	412	25.8	37	4.3	452	15.0
Malaysia	49	2.3	442	8.0	19	2.5	419	14.5	32	2.7	422	10.8
Malta	37	3.0	419	10.1	19	2.2	418	16.7	44	3.4	449	9.0
Mongolia	50	2.3	391	7.4	18	1.7	369	10.1	32	1.8	397	7.0
Morocco	41	5.1	316	9.0	18	2.6	317	20.0	41	5.7	330	11.7
Norway	21	2.1	429	12.6	25	2.5	428	9.8	54	2.8	458	6.1

(continued)

Table 4B.2 (continued)

Country	Older than 10 years				5–10 years old				Younger than 5 years			
	perc.	SE	Math Ach	SE	perc.	SE	Math Ach	SE	perc.	SE	Math Ach	SE
Oman	33	2.2	298	9.3	25	2.0	288	10.1	42	2.8	344	9.0
Palestinian National Authority	26	1.8	314	11.6	31	2.1	335	12.9	43	2.1	339	7.6
Qatar	27	1.0	280	5.3	24	1.1	294	5.6	49	1.1	305	4.2
Romania	38	3.9	362	17.3	21	4.1	340	30.2	41	5.1	420	16.1
Russian Federation	15	2.8	486	20.1	28	4.0	486	12.1	57	4.1	511	9.0
Saudi Arabia	33	2.1	280	10.8	24	1.8	316	9.7	43	2.1	303	7.9
Scotland	32	3.2	436	15.0	19	3.0	452	18.4	49	3.7	454	12.1
Serbia	19	3.3	436	20.1	23	3.5	424	16.5	57	4.1	477	11.7
Singapore	38	2.5	631	6.3	24	2.1	609	9.3	38	2.4	624	9.3
Slovenia	52	4.0	440	11.0	12	2.3	441	21.7	36	3.7	456	10.4
Spain (Basque Country)	35	4.5	437	13.2	38	5.1	457	13.7	27	3.7	450	15.5
Sweden	33	3.3	442	9.3	25	2.5	463	10.3	43	2.8	466	7.9
Syria, Arab Republic of	35	1.8	354	8.0	20	1.4	359	8.2	44	2.0	370	6.2
Thailand	37	10.0	373	36.0	26	8.9	407	78.3	37	10.3	385	45.1
Tunisia	22	3.7	401	11.4	20	3.6	415	17.5	58	5.1	414	9.4
Turkey	27	5.1	460	30.4	33	6.1	377	35.9	40	6.5	394	27.2
Ukraine	23	3.0	371	18.2	19	2.6	382	16.4	58	2.9	411	9.2
United Arab Emirates (Dubai)	35	2.0	492	5.5	23	1.2	489	6.7	42	1.9	486	4.9
United States	19	1.4	459	9.3	32	1.9	480	6.8	49	2.3	465	6.0
United States (Massachusetts)	21	2.9	469	18.3	39	3.3	512	10.4	41	4.6	494	13.2
United States (Minnesota)	30	5.9	453	8.4	29	6.0	484	15.7	42	7.3	510	16.3

Differences Between Girls and Boys

Having examined the age of the students, I will now turn to the other obvious variable, the sex of the students. As stated in Chap. 2, differences between girls and boys are not only a permanent topic in educational research but also a matter of change, especially in recent years where more attention is paid to social justice and adequate schooling for boys and girls (see Chap. 1).

First I analyze the participation of boys and girls in the school system. Table 4B.3 shows the percentage of girls in the school system for first-generation immigrants, native students, and second-generation immigrants. Only the percentage of girls is shown because the percentage of boys is simply 100 minus the percentage of the girls since the missing values are not considered in the analysis. It is also

Table 4B.3 Percentage of girls in schools by different immigration groups

Country	First generation immigrant			Native			Second generation immigrant		
	Perc.	SE	sig	Perc.	SE	sig	Perc.	SE	sig
Armenia	42	3.1	⇩	52	1.0	⇧	52	3.1	
Australia	46	4.1		50	1.7		47	3.1	
Bahrain	30	2.1	⇩	52	0.7	⇧	58	2.3	⇧
Bosnia and Herzegovina	44	1.8	⇩	51	0.9		50	4.4	
Botswana	44	3.7	⇩	54	0.8	⇧	47	2.8	
Bulgaria	34	3.5	⇩	52	1.3	⇧	56	5.6	
Canada (British Columbia)	49	3.3		52	1.3	⇧	51	1.5	
Canada (Ontario)	46	2.7	⇩	52	1.4		49	2.0	
Canada (Quebec)	48	3.7		49	1.5		53	3.3	
Chinese Taipei	31	2.7	⇩	50	1.4		39	5.6	⇩
Colombia	45	3.7	⇩	51	1.6		48	6.2	
Cyprus	50	2.4		50	0.7		49	2.7	
Czech Republic	52	5.1		48	0.9	⇩	50	3.6	
Egypt	45	3.6	⇩	53	2.8		51	4.4	
El Salvador	46	4.1		53	1.5	⇧	49	5.3	
England	46	3.4		52	1.9		49	4.6	
Georgia	30	4.8	⇩	52	1.1	⇧	37	3.5	⇩
Ghana	47	2.0	⇩	45	0.9	⇩	49	3.8	
Hong Kong, SAR	49	2.0		49	1.8		51	1.8	
Hungary	37	5.8	⇩	50	1.2		50	4.8	
Indonesia	46	2.5	⇩	52	1.2	⇧	51	8.1	
Iran, Islamic Republic of	26	9.1	⇩	46	1.5	⇩	44	6.2	
Israel	49	3.2		55	1.8	⇧	55	2.4	⇧
Italy	47	3.3		48	0.7	⇩	49	2.9	
Japan	35	7.7	⇩	51	1.0		41	6.2	
Jordan	37	4.7	⇩	50	1.9		49	3.7	
Korea, Republic of	50	9.2		48	2.7		42	14.2	
Kuwait	39	3.1	⇩	59	2.4	⇧	52	3.4	
Lebanon	48	3.0		57	1.9	⇧	46	4.4	
Lithuania	29	4.1	⇩	51	1.1		57	3.9	⇧
Malaysia	40	2.8	⇩	54	1.5	⇧	57	4.7	⇧
Malta	42	2.6	⇩	51	0.5	⇧	54	2.2	⇧
Mongolia	45	1.9	⇩	53	1.4	⇧	50	2.6	
Morocco	38	4.6	⇩	56	1.7	⇧	33	4.8	⇩
Norway	47	2.9		50	0.8		51	2.3	
Oman	33	4.0	⇩	56	2.1	⇧	50	3.4	
Palestinian National Authority	32	2.2	⇩	56	1.7	⇧	55	3.6	⇧
Qatar	43	0.9	⇩	54	0.7	⇧	54	1.1	⇧
Romania	39	4.8	⇩	50	0.9		29	10.0	⇩

(continued)

Table 4B.3 (continued)

Country	First generation immigrant			Native			Second generation immigrant		
	Perc.	SE	sig	Perc.	SE	sig	Perc.	SE	sig
Russian Federation	51	3.5		52	0.9	⇧	54	2.4	⇧
Saudi Arabia	28	2.8	⇩	53	2.3		50	3.8	
Scotland	42	3.1	⇩	52	1.1	⇧	53	3.5	
Serbia	50	3.9		50	0.8		48	2.0	
Singapore	50	2.8		48	1.0	⇩	50	1.7	
Slovenia	33	3.7	⇩	52	0.9	⇧	49	1.9	
Spain (Basque Country)	41	3.3	⇩	49	1.7		48	5.5	
Sweden	40	2.7	⇩	48	1.0	⇩	51	2.0	
Syria, Arab Republic of	46	2.7		55	2.0	⇧	48	3.9	
Thailand	36	10.4		50	1.3		38	5.3	⇩
Tunisia	34	3.2	⇩	53	0.9	⇧	45	4.5	
Turkey	44	6.7		47	0.8	⇩	43	5.1	⇩
Ukraine	39	2.8	⇩	54	1.0	⇧	52	1.7	
United Arab Emirates (Dubai)	47	5.8		57	4.8	⇧	52	6.5	
United States	47	2.2	⇩	51	0.8	⇧	52	1.6	
United States (Massachusetts)	46	3.3		50	1.3		51	3.7	
United States (Minnesota)	46	4.9		54	1.3	⇧	44	4.7	

indicated that the percentage of girls is significantly higher than the percentage of boys for each group. However, we need to be cautious here: due to the huge sample sizes, very small deviations appear as statistically significant, i.e., even less than a 1 % difference can be statistically significantly different.

Looking first at the group of native students, I gather that in several Arabic countries, girls are overrepresented in the schools. In Kuwait 59 % of native students are girls and consequently only 41 % are boys. In Dubai I find 57 % girls and 56 % in Lebanon. In Oman, Morocco, and Palestine, there are 56 % girls and in Syria 55 %. Also Israel, which has a certain percentage of Arabic population, has 55 % girls in the school population. Interestingly, in Iran, which is a neighboring country to the Arabic countries and shares with them the Islamic religion, the case is different with 54 % boys and only 46 % girls in the school system. Further interesting cases are Ghana and Botswana. In Botswana I find 54 % girls and 46 % boys; in Ghana there are 45 % girls and 55 % boys. For most other countries, the percentages of boys and girls are within a plus/minus 2 % range around the fifty-fifty.

For the immigrant populations, we see much more fluctuation. In several countries I observe a relatively low percentage of first-generation immigrant girls. In Iran only 26 % of the first-generation immigrant students are girls and in Saudi Arabia only 28 %. Other countries with less than 40 % of girls in this group are Bahrain, Bulgaria, Chinese Taipei, Georgia, Hungary, Japan, Jordan, Kuwait, Lithuania, Morocco, Oman, Palestine, Romania, Slovenia, Thailand, and Tunisia.

This is a very surprising result and definitely a matter of policy concern. The list of countries where the participation of first-generation immigrant girls in the school system is seriously lower than that of boys includes developing as well as developed countries in all regions of the world. Further research is necessary to uncover if these girls are taught differently – for example, by private tutors and why they are not integrated in the school systems of the respective host country. Whatever the reasons, we observe that in 34 out of 56 countries, there are statistically significantly fewer first-generation immigrant girls enrolled in the education system in grade eight and there is not a single country where first-generation immigrant boys are underrepresented in the education system according to TIMSS 2007 data. I can probably conclude that first-generation girls are discriminated and their chances for integration in the host countries are reduced. Policy implications could be to focus on this group and implement initiatives to include them into the educational system.

For second-generation immigrant students, the picture is much more diverse. I can find countries where girls are underrepresented in the educational system – for example, Georgia, Chinese Taipei, Morocco, or Thailand. But I also find countries where boys are underrepresented – for example, Bahrain, Lithuania, or Malaysia; however, this underrepresentation of boys is in general less pronounced than that of girls.

I conclude that there are substantial differences in the participation of immigrant boys and girls in the educational systems. Mostly, I find countries where girls are underrepresented in the educational system.

Inclusion into the educational system is one aspect, but the achievement of immigrant boys and girls in the educational system also needs some evaluation.

Table 4B.4 displays the mathematics achievement of grade eight boys and girls in TIMSS 2007 separately for immigrant boys and girls. As for the percentages of boys and girls in the system, I can also observe substantial differences regarding the achievement of boys and girls that participate in the educational systems. Although in most countries first-generation and second-generation immigrant students are lagging behind independent of their sex, I find quite a number of countries where the difference is less pronounced for girls. In Iran – a country with significantly lower participation of first-generation immigrant girls – first-generation immigrant boys have 47 score points behind native boys, whereas first-generation immigrant girls have 28 score points ahead of native girls. So, first-generation immigrant girls that are included in the educational system are performing quite well and much better than their male peers. In Japan I detect a similar pattern of lower participation of girls but higher achievement of first-generation immigrant girls compared with their male peers.

Turkey is also a very interesting case where the participation of girls was slightly lower – but not statistically significant – but first-generation immigrant boys are lagging behind native boys by 52 score points, whereas first-generation immigrant girls are at about the same level as native girls and boys. On average first-generation immigrant boys were lagging behind native boys by 39 score points and first-generation immigrant girls by 29 score points.

Table 4B.4 Mathematics achievement of immigrant boys and girls

Country	First generation immigrant				Native				Second generation immigrant			
	Boy		Girl		Boy		Girl		Boy		Girl	
	Math Ach	SE	Math Ach	SE	Math Ach	SE	Math Ach	SE	Math Ach	SE	Math Ach	SE
Armenia	501	12.2	515	18.5	497	3.5	500	3.3	501	12.3	506	15.6
Australia	505	10.0	487	11.1	503	5.4	488	3.9	508	7.6	492	11.9
Bahrain	362	5.6	413	6.0	390	2.6	416	2.6	401	6.3	410	4.4
Bosnia and Herzegovina	449	5.2	457	4.9	458	2.8	456	3.7	460	11.0	464	8.4
Botswana	334	13.4	364	12.9	362	3.4	374	2.4	330	6.0	353	6.7
Bulgaria	409	10.6	406	13.6	467	6.2	478	4.6	444	24.9	476	21.4
Canada (British Columbia)	535	7.7	533	9.1	501	3.4	497	3.0	523	4.9	511	3.9
Canada (Ontario)	543	7.4	523	8.4	515	5.0	510	5.3	525	5.1	517	4.7
Canada (Quebec)	514	12.1	514	8.0	532	3.9	531	3.5	533	9.9	527	8.0
Chinese Taipei	498	12.3	489	15.3	608	5.0	604	4.5	600	17.6	578	16.4
Colombia	325	9.4	309	17.3	402	3.9	369	4.0	367	18.4	319	16.0
Cyprus	417	6.9	439	7.8	460	2.7	482	2.1	464	6.5	467	6.4
Czech Republic	480	12.5	493	8.4	504	2.9	507	2.6	493	7.1	491	9.0
Egypt	347	5.0	357	7.6	426	5.1	428	4.6	351	9.7	369	12.8
El Salvador	311	12.3	280	13.1	355	4.0	335	3.9	352	10.3	330	10.7
England	499	11.9	490	12.9	519	6.4	511	5.2	528	7.5	528	8.6
Georgia	371	15.1	371	14.8	418	6.9	417	6.0	356	14.9	374	25.5
Ghana	271	6.9	256	6.4	334	4.5	311	5.0	308	13.2	292	11.8
Hong Kong, SAR	546	11.6	560	7.4	573	8.5	584	5.3	579	7.7	586	6.1
Hungary	459	22.9	463	26.5	519	3.4	519	4.0	538	14.7	508	14.7
Indonesia	353	6.6	349	7.2	406	4.7	412	4.1	357	18.9	340	28.2
Iran, Islamic Republic of	355	18.6	436	34.5	402	6.1	408	5.3	351	13.4	386	18.7
Israel	437	7.7	447	12.3	471	5.4	468	5.0	486	8.1	475	6.4
Italy	455	8.3	447	7.6	485	3.6	477	3.6	475	8.6	486	7.1
Japan	519	24.9	569	33.6	573	3.4	569	3.1	559	15.3	562	21.0
Jordan	376	13.1	399	12.2	420	6.5	439	6.7	446	6.7	455	7.4
Korea, Republic of	655	26.9	612	43.0	600	3.0	595	3.3	523	37.5	543	57.1
Kuwait	313	8.6	352	7.0	356	4.4	366	3.0	352	8.7	374	4.5
Lebanon	429	6.8	416	6.6	469	4.7	451	4.3	449	11.2	457	9.4
Lithuania	439	8.8	461	19.0	508	2.4	511	3.1	508	10.0	509	6.2
Malaysia	422	9.3	439	9.2	475	5.0	483	5.6	438	15.0	462	9.9
Malta	424	9.0	442	8.4	495	2.0	492	1.9	494	5.1	489	5.4
Mongolia	393	6.6	385	7.2	453	4.4	443	4.0	399	5.2	393	6.2
Morocco	326	8.5	318	10.4	396	4.8	381	3.7	362	11.6	357	12.6
Norway	442	5.8	447	6.6	473	2.9	475	2.3	459	5.5	465	5.3

(continued)

Table 4B.4 (continued)

Country	First generation immigrant				Native				Second generation immigrant			
	Boy		Girl		Boy		Girl		Boy		Girl	
	Math Ach	SE	Math Ach	SE	Math Ach	SE	Math Ach	SE	Math Ach	SE	Math Ach	SE
Oman	296	7.0	358	8.8	362	5.3	406	3.7	348	7.6	385	6.5
Palestinian National Authority	317	9.0	354	10.4	367	5.3	390	4.3	341	12.1	389	8.8
Qatar	274	3.2	323	5.2	287	3.0	319	2.7	317	4.2	341	3.3
Romania	379	12.9	386	20.5	459	4.7	474	4.1	337	36.7	458	13.6
Russian Federation	495	11.3	504	9.2	514	4.7	516	4.1	501	8.6	505	9.5
Saudi Arabia	289	7.6	320	9.1	327	3.9	342	4.0	342	8.2	359	6.5
Scotland	448	12.1	453	13.6	493	4.2	489	3.7	505	9.1	494	9.0
Serbia	445	12.7	468	10.2	486	4.3	490	3.9	490	5.7	501	6.2
Singapore	610	8.2	634	6.2	581	4.8	596	4.1	594	7.2	600	6.8
Slovenia	441	8.5	461	12.3	512	3.0	505	2.8	491	6.8	484	5.5
Spain (Basque Country)	446	11.7	451	10.0	508	3.7	502	3.6	498	12.4	485	11.7
Sweden	452	7.1	466	8.6	497	2.5	500	2.6	485	5.1	480	4.0
Syria, Arab Republic of	370	7.7	357	7.0	422	4.5	398	3.9	385	10.1	387	8.8
Thailand	351	48.9	433	26.7	432	5.5	454	5.3	395	20.7	448	19.2
Tunisia	414	8.0	391	10.0	435	2.6	412	2.8	398	8.8	389	9.0
Turkey	383	17.8	432	32.2	434	5.1	433	5.4	398	26.5	410	21.2
Ukraine	396	9.6	400	10.2	465	3.9	467	3.9	479	6.0	478	5.6
United Arab Emirates (Dubai)	490	7.8	486	6.3	392	10.3	401	7.8	469	6.5	456	7.0
United States	469	5.9	467	6.5	520	3.0	514	3.0	499	5.1	497	4.6
United States (Massachusetts)	498	11.1	495	14.4	560	5.1	555	4.2	540	8.9	525	12.5
United States (Minnesota)	483	13.0	489	15.4	542	5.1	536	4.2	519	11.6	510	8.1

In 42 out of 56 countries, first-generation immigrant boys were performing statistically significantly lower than native boys. First-generation immigrant girls performed statistically significantly lower than native girls in 35 countries. In contrast to this, in five countries first-generation immigrant boys performed statistically significantly better than native boys, and in three countries first-generation immigrant girls outperformed native girls statistically significantly.

For second-generation immigrants, the situation is similar although not that extreme. On average boys in this group are lagging behind native boys by 14 score points and girls are outperformed by ten score points. A very pronounced case is that of Romania where second-generation immigrant boys are lagging behind their native peers by 122 score points and second-generation immigrant girls are lagging behind native girls by 15 score points, which is even not statistically significant.

Second-generation immigrant boys are performing statistically significantly lower than native boys in 16 out of 56 countries. Second-generation immigrant girls perform statistically significantly lower than native girls in 13 countries. Just as for first-generation immigrant boys, I find five countries where second-generation immigrant boys performed statistically significantly better than native boys, but four countries where second-generation immigrant girls outperformed native girls statistically significantly.

In order to achieve improved situations for immigrant children, policy makers need to address the issues of boys and girls differently in a good number of countries. In some countries the major requirement for immigrant girls seems to be their inclusion into the educational system. For immigrant boys the emphasis should be more on the educational outcome. But these hypotheses need further research before final conclusions can be drawn and policy recommendations can be given.

I conclude that the situation is different for boys and girls in a number of countries. Especially for first-generation immigrant students, I find fewer girls enrolled in the schools in the majority of countries. For first-generation immigrant boys, I find that they are lagging behind in mathematics more often and to a greater extent than first-generation immigrant girls. For second-generation immigrant boys and girls, the difference is less pronounced, but also in this group, immigrant boys tend to lag behind native boys more often and to a greater extent.

Interestingly, in the Canadian provinces of British Columbia and Ontario as well as in Singapore, I could not observe that immigrant girls are underrepresented in the schools or that they perform less well than immigrant boys.

Language Spoken by Immigrant Students

I now address the students' home background starting with the language spoken at home. Language difficulties are often considered as the main factor for the lower performance of immigrant students by researchers (see, e.g., Buchmann and Parrado (2006); see my review in Chap. 2).

In TIMSS the students were asked how often they speak the language of the test at home (Fig. 4B.2).

Table 4B.5 lists the percentage of students who answered that they never or only sometimes speak the language of test at home separately for native students and first- and second-generation immigrant students. It should be noted again that the TIMSS test design requested the countries to administer the test in the language of instruction used in the schools. Consequently, students who do not speak the language of the test at home are students who do not speak the language of instruction used in the schools at home.

In 32 out of 56 countries, the percentage of first-generation immigrant students that do not speak the language of the test is significantly higher than the percentage of native students that do speak the language of the test. Interestingly, I find seven countries with a smaller percentage of first-generation immigrant students than

3 ▰▰▰▰▰▰▰▰▰▰▰▰▰▰▰▰▰▰▰▰▰▰▰▰▰▰▰▰▰▰▰▰

How often do you speak <language of test> at home?

*Fill in **one** circle only*

Always ------------------------------------- ①

Almost always ----------------------------- ②

Sometimes ------------------------------- ③

Never ------------------------------------- ④

Fig. 4B.2 TIMSS 2007 student questionnaire, variable BS4GOLAN (Source: TIMSS 2007 Assessment. Copyright © 2009 International Association for the Evaluation of Educational Achievement (IEA). Publisher: TIMSS & PIRLS International Study Center, Lynch School of Education, Boston College)

native students who never or only sometimes speak the language of instruction at home, namely the Arabic countries Qatar, Tunisia, Kuwait, and Lebanon; the African countries Ghana and Botswana; and Malta.

Table 4A.3 displays the achievement results for the countries with a higher percentage of native students that never or only sometimes speak the language of instruction at home. None of the countries obtained positive mathematics achievement results for immigrant students compared to native students with the exception of Botswana where the achievement difference between native students and first-generation immigrant students was not significant. In all other countries, immigrant students had a statistically significantly lower mathematics achievement than native students. In Ghana first-generation immigrant students achieved 60 score points less than native students; in Malta the difference was even 62 points. On the other hand, all of the participants whose first-generation immigrant students outperform their native students have a statistically significantly higher percentage of first-generation immigrant students that speak the language of the test never or only sometimes at home compared to native students (see Table 4B.5). These participants are Australia, the Canadian provinces of British Columbia and Ontario, Singapore, and Dubai.

Twenty-five countries show a higher percentage of second-generation immigrant students than native students that speak the language of instruction never or only sometimes at home. Again, I find examples where the percentage of students that never or only sometimes speak the language of instruction at home is lower than among their native peers. This is the case in nine countries, among them, again, Arabic countries like Oman, Saudi Arabia, Morocco, Qatar, Kuwait, and Lebanon as well as Malaysia, Botswana, and Malta.

I am now going to analyze the achievement of students who never or only sometimes speak the language of the test at home, focusing on the achievement difference for students who speak the language of the test at home and those who don't. Table 4B.6 displays the mathematics achievement of the students who, in TIMSS

Table 4B.5 Percentages of students who do not speak the language of the test at home

Country	1st gen not speaking the language			Natives not speaking the language		2nd gen not speaking the language		
	Percent	SE	sig	Percent	SE	Percent	SE	sig
United States (Massachusetts)	44	3.7	⇧	1	0.4	17	2.7	⇧
Norway	38	2.9	⇧	0	0.1	14	1.9	⇧
United States (Minnesota)	37	4.8	⇧	1	0.2	17	3.6	⇧
Czech Republic	37	5.3	⇧	1	0.3	5	1.3	⇧
Canada (British Columbia)	42	2.3	⇧	6	2.1	15	2.2	⇧
United States	36	2.6	⇧	2	0.3	24	1.6	⇧
Sweden	33	2.8	⇧	0	0.1	15	1.8	⇧
Canada (Ontario)	37	3.9	⇧	4	1.4	10	1.4	⇧
Canada (Quebec)	35	3.4	⇧	3	0.5	23	2.9	⇧
Israel	33	2.7	⇧	2	0.3	7	1.0	⇧
Japan	29	8.1	⇧	1	0.2	5	3.0	⇧
Iran, Islamic Republic of	63	10.8	⇧	37	2.2	34	6.7	
Slovenia	32	3.8	⇧	7	1.1	20	2.8	⇧
England	23	3.0	⇧	1	0.2	5	1.4	⇧
Thailand	51	10.6		33	1.9	33	10.6	
Australia	18	2.1	⇧	1	0.2	7	1.2	⇧
Scotland	20	3.1	⇧	3	0.4	6	1.7	⇧
Singapore	67	2.5	⇧	51	1.0	56	1.8	⇧
Cyprus	21	2.0	⇧	6	0.5	16	1.5	⇧
Malaysia	51	4.2	⇧	35	2.1	29	4.2	⇩
Hungary	16	5.8	⇧	1	0.2	5	2.5	⇧
Italy	12	2.3		0	-	3	1.0	
Bulgaria	19	3.2	⇧	10	1.6	11	3.9	
El Salvador	10	2.3	⇧	3	0.3	3	1.4	
Spain (Basque Country)	14	2.6	⇧	6	0.5	8	2.4	
United Arab Emirates (Dubai)	44	1.5	⇧	37	2.8	38	2.7	
Lithuania	8	2.3	⇧	1	0.4	7	1.7	⇧
Georgia	11	4.7		5	0.8	14	4.0	⇧
Palestinian National Authority	17	2.0	⇧	11	1.5	10	2.8	
Russian Federation	12	4.1		6	1.8	6	2.5	
Hong Kong, SAR	13	2.7	⇧	8	0.6	8	1.0	
Colombia	9	2.4	⇧	4	0.4	7	2.5	⇧
Syria, Arab Republic of	18	1.5	⇧	13	1.1	13	2.5	
Mongolia	8	1.6	⇧	4	0.5	6	1.2	⇧
Bahrain	22	1.9	⇧	17	1.0	23	1.7	⇧

(continued)

Table 4B.5 (continued)

Country	1st gen not speaking the language			Natives not speaking the language		2nd gen not speaking the language		
	Percent	SE	sig	Percent	SE	Percent	SE	sig
Chinese Taipei	21	3.4		16	1.2	15	3.5	
Ukraine	33	3.7		30	2.7	38	3.8	⇧
Bosnia and Herzegovina	4	0.8	⇧	1	0.4	2	1.0	
Serbia	5	2.1		2	0.7	2	1.1	
Jordan	12	1.3		10	1.0	11	1.4	
Armenia	5	1.1		2	0.5	6	1.8	⇧
Turkey	12	3.9		10	1.2	9	3.0	
Romania	3	1.2		2	0.3	0	-	
Egypt	17	1.3		16	1.2	14	3.2	
Indonesia	62	4.2		64	2.9	57	9.8	
Oman	22	2.3		25	2.3	19	2.4	⇩
Saudi Arabia	26	2.6		29	2.6	22	2.3	⇩
Morocco	45	5.9		48	1.8	37	4.2	⇩
Qatar	25	0.9	⇩	29	0.7	27	0.9	⇩
Korea, Republic of	0	-		5	0.4	0	-	
Tunisia	72	2.9	⇩	78	1.0	75	3.8	
Kuwait	27	1.9	⇩	35	1.5	30	2.3	⇩
Ghana	61	2.0	⇩	70	1.3	67	3.9	
Lebanon	70	2.9	⇩	81	1.2	78	3.5	⇩
Botswana	50	4.6	⇩	67	1.1	57	2.1	⇩
Malta	66	2.6	⇩	87	0.5	67	2.0	⇩

2007, answered that they speak the language of the test never or sometimes at home (group 1) and those who answered that they speak the language of the test always or almost always at home (group 2). The achievement difference is calculated together with an indicator of its statistical significance. The results are displayed separately for native students and for first- and second-generation immigrant students.

For native students I observe that in 29 countries, the mathematics achievement of the students in group 1 is statistically significantly lower than that of the students in group 2. Interestingly, in Egypt, Malaysia, Tunisia, and Dubai, the opposite is the case and the students in group 2 achieve statistically significantly higher than the students in group 1.

For first-generation immigrant students, there are only 11 countries where the mathematics achievement of the students in group 1 is statistically significantly lower than that of the students in group 2. In Bahrain and Tunisia, the achievement of the former is statistically significantly higher than that of the latter.

The smaller number of countries with a statistically significantly higher mathematics achievement of group 2 students is of course partially an effect of the group

Table 4B.6 Mathematics achievement of native and immigrant students speaking or not speaking the language of the test of home

Country	First generation immigrant							Native							Second generation immigrant						
	Speaking language at home		Not speaking language at home		Difference			Speaking language at home		Not speaking language at home		Difference			Speaking language at home		Not speaking language at home		Difference		
	Math Ach	SE	Math Ach	SE				Math Ach	SE	Math Ach	SE				Math Ach	SE	Math Ach	SE			
Armenia	506	13.3	525	23.9	-19.0			499	3.0	472	10.6	26.4	⇧		502	12.0	523	22.1	-20.8		
Australia	500	6.9	484	17.2	15.6			496	3.8	419	25.5	77.5	⇧		500	6.1	500	22.8	0.4		
Bahrain	372	4.5	395	7.8	-23.1	⇩		403	2.2	407	4.9	-4.5			402	3.8	421	9.1	-18.2		
Bosnia and Herzegovina	454	4.6	422	15.1	31.8	⇧		457	3.0	442	19.6	15.1			461	7.5	498	42.1	-36.6		
Botswana	379	12.7	316	12.6	62.9	⇧		370	3.2	368	2.5	2.5			353	7.1	332	5.5	21.6		
Bulgaria	414	10.1	380	20.9	34.6			479	4.5	414	15.2	65.0	⇧		465	18.3	439	55.6	26.4		
Canada (British Columbia)	526	7.8	546	9.4	-19.9			498	2.8	511	6.6	-12.5			518	3.8	514	7.0	3.5		
Canada (Ontario)	528	6.7	544	11.6	-16.5			515	3.7	450	25.9	65.4	⇧		520	4.1	529	8.6	-8.8		
Canada (Quebec)	510	8.0	521	13.8	-11.3			531	3.2	527	13.2	4.8			529	6.6	532	14.2	-3.1		
Chinese Taipei	511	10.4	434	19.1	77.0	⇧		616	4.1	556	7.6	60.1	⇧		603	12.5	528	37.3	74.3		
Colombia	320	8.1	296	16.1	24.2			386	3.6	346	6.9	40.4	⇧		344	14.4	341	21.4	3.0		
Cyprus	432	5.8	415	12.6	16.9			474	1.8	438	7.1	35.3	⇧		467	5.4	457	11.2	10.1		
Czech Republic	477	7.6	503	13.9	-25.4			506	2.5	466	14.8	39.7	⇧		492	5.6	492	18.7	-0.7		
Egypt	349	5.4	363	8.8	-14.0			425	3.5	440	6.2	-15.1	⇩		357	9.3	378	21.7	-20.9		
El Salvador	298	11.3	286	19.4	12.6			346	2.9	294	9.7	51.1	⇧		343	7.7	288	22.7	54.8	⇧	
England	489	10.0	514	21.7	-24.8			515	5.1	465	18.8	49.4	⇧		526	6.2	566	14.6	-40.4	⇩	
Georgia	367	12.2	404	57.5	-37.2			418	6.0	408	21.2	10.2			366	16.6	344	28.2	21.2		
Ghana	254	6.4	270	6.8	-16.3			327	5.7	323	4.3	4.1			301	17.0	299	10.3	1.6		
Hong Kong, SAR	566	6.0	468	16.4	97.5	⇧		582	5.9	535	13.2	47.6	⇧		588	5.8	516	17.1	72.5	⇧	

(continued)

Table 4B.6 (continued)

	First generation immigrant					Native					Second generation immigrant				
	Speaking language at home		Not speaking language at home		Difference	Speaking language at home		Not speaking language at home		Difference	Speaking language at home		Not speaking language at home		Difference
Country	Math Ach	SE	Math Ach	SE		Math Ach	SE	Math Ach	SE		Math Ach	SE	Math Ach	SE	
Hungary	463	16.1	451	51.5	11.1	519	3.3	476	32.2	43.8	524	11.1	508	32.3	15.8
Indonesia	350	8.8	352	6.1	-2.5	409	6.0	409	4.5	-0.4	351	30.9	346	17.4	5.1
Iran, Islamic Republic of	395	33.8	364	22.8	30.7	424	4.8	372	4.4	51.6 ⇦	388	12.8	325	17.8	62.8 ⇦
Israel	441	9.6	445	10.9	-4.3	471	4.1	411	19.6	60.1 ⇦	481	6.0	465	19.4	15.6
Italy	452	6.8	445	14.6	7.3	481	3.2			⇦	482	5.9	427	38.8	55.3
Japan	560	22.8	478	28.0	81.9	572	2.5	527	11.0	44.4 ⇦	568	13.6	418	74.3	150.5 ⇦
Jordan	385	8.1	384	19.3	1.1	431	4.6	416	11.0	14.8	451	4.8	445	14.1	5.5
Korea, Republic of	633	28.4				600	2.7	551	7.8	49.1 ⇦	531	33.7			⇦
Kuwait	328	6.6	329	9.8	-1.3	364	2.9	358	4.4	5.9	364	5.8	362	7.6	1.7
Lebanon	428	10.7	421	5.4	7.3	466	7.6	457	3.9	9.3	473	11.2	447	8.9	25.3
Lithuania	444	9.1	463	20.1	-19.7	510	2.3	476	12.5	33.9	507	5.7	538	16.3	-31.2 ⇦
Malaysia	421	8.8	437	13.0	-16.2	469	5.5	498	6.9	-28.5 ⇦	441	11.8	477	12.7	-36.2 ⇨
Malta	476	9.8	407	7.6	68.9 ⇦	516	4.2	490	1.7	25.3 ⇦	490	5.7	492	5.2	-2.1 ⇦
Mongolia	393	5.7	348	14.1	45.4 ⇦	449	3.8	411	13.9	37.7	398	4.1	366	20.5	31.0
Morocco	322	7.9	324	11.7	-2.5	383	4.1	393	4.3	-9.4	352	12.4	373	10.8	-21.0 ⇦
Norway	451	5.6	435	7.2	15.5	474	2.2	445	25.0	29.1	467	3.9	429	11.2	38.8
Oman	316	6.6	318	9.5	-1.6	386	3.5	389	5.1	-3.4	365	7.0	373	11.2	-7.6
Palestinian National Authority	331	7.7	321	13.4	9.2	380	3.7	382	7.8	-2.5	365	8.2	389	21.9	-23.8
Qatar	302	3.3	275	6.9	26.8 ⇦	308	2.2	296	3.4	12.1 ⇦	335	3.0	317	5.6	18.0 ⇦

Country																		
Romania	381	11.8	399	62.9	-18.1		467	4.1	408	15.1	58.9	⇦	372	30.4	452		372.2	⇦
Russian Federation	499	8.0	506	30.2	-6.9		516	3.8	501	12.8	14.9		506	6.4	359	20.5	54.7	⇦
Saudi Arabia	299	6.2	295	10.1	3.5		333	3.5	339	4.3	-6.0		348	5.7	481	9.7	-10.4	
Scotland	445	11.7	471	15.2	-25.5		493	3.5	432	9.9	60.9	⇦	500	7.3	481	21.4	19.3	⇦
Serbia	457	9.6	437	40.8	20.1		489	3.7	426	17.0	63.5	⇦	497	4.3	420	22.3	76.8	⇦
Singapore	632	7.6	617	7.3	15.2		614		563	4.8	50.3	⇦	621	6.2	578	7.4	43.4	⇦
Slovenia	454	8.1	434	12.6	20.3		511	2.1	474	8.3	37.0	⇦	495	4.3	458	8.9	37.4	⇦
Spain (Basque Country)	446	9.2	461	16.5	-15.2		505	2.9	505	5.2	0.1		490	9.3	509	20.7	-18.5	
Sweden	460	6.3	453	10.2	7.3		499	2.2	455	18.6	43.5	⇦	485	3.3	469	9.1	15.9	
Syria, Arab Republic of	365	5.6	360	11.4	4.8		410	3.6	400	5.6	9.4		385	7.1	395	17.6	-10.1	
Thailand	438	39.5	327	34.3	110.6	⇦	458	5.9	413	7.7	44.5	⇦	427	17.6	392	50.9	34.6	
Tunisia	388	10.8	413	6.8	-25.6	⇨	410	3.8	426	2.5	-16.6	⇨	373	10.9	401	7.2	-27.7	⇨
Turkey	414	16.8	334	34.1	80.4	⇦	441	5.0	370	6.0	71.1	⇦	413	20.4	297	31.3	116.4	⇦
Ukraine	391	9.8	409	13.1	-18.0		465	4.4	469	4.5	-3.8		473	6.4	487	6.5	-13.8	
United Arab Emirates (Dubai)	496	4.7	478	5.1	18.9	⇦	377	6.6	431	9.0	-53.6	⇨	460	3.9	465	6.4	-4.4	
United States	467	5.7	470	7.1	-2.5		518	2.8	457	6.4	61.4	⇦	504	4.3	479	6.1	25.2	⇦
United States (Massachusetts)	504	10.3	488	14.4	16.0		558	4.0	528	16.6	29.5		542	8.5	488	13.0	54.0	⇦
United States (Minnesota)	492	10.6	476	15.9	16.6		539	4.1	487	26.1	52.1	⇦	516	7.5	510	11.7	6.1	

sizes. Since there are more native students and fewer first-generation immigrant students in the sample, the sampling errors for the subgroups within first-generation immigrant students are greater than for native students. This can also be seen in the columns labeled "SE" in Table 4B.6. This, however, is not the only reason. The average achievement difference between the students in the two groups is 26.7 score points for native students and only 14 score points for first-generation immigrant students. Based on this data, one might conclude that the effect of not speaking the language of the test (and consequently the language of school instruction) at home makes a bigger difference for native students than for first-generation immigrant students.

But we should not forget to take into account the results for second-generation immigrant students. For the second-generation immigrant students, there are 15 countries where the mathematics achievement of the students in group 1 is statistically significantly lower than that of the students in group 2. On the other hand, in England, Malaysia, and Tunisia, the mathematics achievement of the first group is statistically significantly higher than that of the second group. Again, the group sizes for second-generation immigrant students are smaller than for native students, and the error terms are, therefore, greater. However, the difference for second-generation immigrant students is on average smaller than for native students – in this case 15.5 score points.

I conclude that in many countries, there are more students who do not speak the language of instruction at home among immigrant students than among native students. Interestingly, the difference in mathematics achievement for students speaking the language of instruction at home and those who don't is bigger for native students than for first- and second-generation immigrant students.

Parents' Education

There are further factors of the home background that might be related to the students' achievement and that might be different for native and immigrant students. As reflected in Chap. 2, the socioeconomic background of the students plays a major role in predicting the students' achievement. Consequently, the next analysis will investigate the relationship between the socioeconomic status of the students and their immigration status. Not all domains are covered in the analyzed data, for example, there is no indication of symbolic capital. Measures of economic, cultural, and social capital are not operationalized in an optimal way, and variables are measuring a mixture of them. Unlike, e.g., PIRLS that includes parent questionnaires, in TIMSS only information from the students at the individual level is available. Questions that can be used to describe the socioeconomic background of the students are questions about the highest level of education of the parents, the number of books in the home, as well as a set of nine home possessions, five of which are internationally defined (calculator, computer, study desk, dictionary, and Internet connection), while the other four were defined by each participating country considering items that would discriminate well between students from low and high socioeconomic background. However according to Brese and Mirazchiyski (2010), these possession items seem not to work as good as the parental education and the number of books at home.

I will start with the parents' education. TIMSS 2007 includes two questions about the highest level of education that is completed by the mother and the father. The question in the international version of the questionnaire refers to the different ISCED levels. ISCED stands for the "international standard classification of education" and is defined by the UNESCO originally in 1997 (see UNESCO, 2006). In 2011 the classification was revised by the UNESCO, but since TIMSS followed the ISCED definition from 1997, I will follow the 1997 definition in this publication.

ISCED level 2 is defined as the lower secondary or second stage of basic education which is defined by the following criteria: "... entry is after some 6 years of primary education (see paragraph 35; the end of this level is after some 9 years of schooling since the beginning of primary education (see paragraph35); the end of this level often coincides with the end of compulsory education in countries where this exists; and often, at the beginning of this level, several teachers start to conduct classes in their field of specialization" (UNESCO, 2006, p. 24).

ISCED level 5 is defined as the first stage of tertiary education with the criteria: "... normally the minimum entrance requirement to this level is the successful completion of ISCED level 3Aa or 3B or ISCED level 4A; level 5 programmes do not lead directly to the award of an advanced research qualification (level 6); and these programmes must have a cumulative theoretical duration of at least 2 years from the beginning of level 5" (UNESCO, 2006, p. 34).

The participating countries were required to translate the definitions into terms that are familiar to students in their country. The TIMSS 2007 question in the student questionnaire was (Fig. 4B.3):

In the style of Mullis et al. (2004) (see page 126 ff.), the highest level of education of the parents was calculated and the analysis was based on this derived variable.

First, the percentage of students was calculated with at least one of the parents having an education of ISCED level 5 or higher. The percentages were calculated separately for native students and first- and second-generation immigrant students. Table 4B.7 shows these statistics for TIMSS 2007 countries together with an indicator showing the percentage for one of the immigrant population groups was statistically significantly different from the percentage for native students.

We can see in Table 4B.7 that there are more countries (nineteen) with a statistically significantly higher percentage especially of first-generation immigrant students than native students with one of the parents having completed an ISCED level 5 education. Only in seven countries, however, there were statistically significantly more native students than first-generation immigrant students with at least one of their parents having completed an education of ISCED level 5 or higher.

Apart from Korea which has – as stated above – very few first-generation immigrant students and consequently no reliable information, the participants with the biggest differences of the percentages between native students and first-generation immigrant students are exactly those that have shown more positive mathematics achievement results: Singapore, Dubai, and the Canadian provinces of British

6

A. What is the highest level of education completed by your mother (or stepmother or female guardian)?

*Fill in **one** circle only*

Some <ISCED Level 1 or 2 > or did not
go to school ------------------------ ①

<ISCED 2>----------------------- ②

<ISCED 3>----------------------- ③

<ISCED 4>----------------------- ④

<ISCED 5B>--------------------- ⑤

<ISCED 5A, first degree> ------------ ⑥

Beyond<ISCED 5A, first degree> ------- ⑦

I don't know---------------------- ⑧

B. What is the highest level of education completed by your father (or stepfather or male guardian)?

*Fill in **one** circle only*

Some <ISCED Level 1 or 2 > or did not
go to school ------------------------ ①

<ISCED 2>----------------------- ②

<ISCED 3>----------------------- ③

<ISCED 4>----------------------- ④

<ISCED 5B>--------------------- ⑤

<ISCED 5A, first degree> ------------ ⑥

Beyond<ISCED 5A, first degree> ------- ⑦

I don't know---------------------- ⑧

Fig. 4B.3 TIMSS 2007 student questionnaire, variables BS4GMFED and BS4GFMFED (Source: TIMSS 2007 Assessment. Copyright © 2009 International Association for the Evaluation of Educational Achievement (IEA). Publisher: TIMSS & PIRLS International Study Center, Lynch School of Education, Boston College)

Columbia, Quebec, and Ontario. Interestingly, Quebec is among them although the mathematics achievement of immigrant students compared to native students was not as positive as in the other two Canadian provinces. Since I already learned from the literature review as well as from the TIMSS 2007 international results that a higher educational background of the parents and the achievement level of the students are correlated, I could assume that the higher education levels in these countries are contributing to the relatively positive achievement outcomes of immigrant

Table 4B.7 Percentage of students whose parents' highest education is ISCED level 5 and higher

Country	First generation immigrant			Native		Second generation immigrant		
	Percent	SE	sig	Percent	SE	Percent	SE	sig
Korea, Republic of	83	9.6	⇧	52	1.4	11	11.5	⇩
Singapore	64	2.4	⇧	33	1.1	36	1.8	
United Arab Emirates (Dubai)	75	1.6	⇧	46	2.9	56	2.2	⇧
Canada (British Columbia)	76	2.7	⇧	49	2.1	53	2.3	
Canada (Quebec)	71	3.7	⇧	44	1.8	50	3.2	
Canada (Ontario)	73	3.0	⇧	48	2.4	54	3.3	
Cyprus	61	2.5	⇧	41	0.9	61	2.3	⇧
Czech Republic	40	5.3	⇧	24	1.0	36	3.6	⇧
Australia	56	4.5	⇧	41	1.7	42	2.0	
Japan	58	11.3		44	1.3	69	7.7	⇧
Palestinian National Authority	36	1.8		23	1.1	26	2.7	
Bahrain	36	2.1	⇧	24	0.9	22	2.7	
Morocco	33	3.8	⇧	21	1.5	24	3.7	
Italy	40	4.7	⇧	29	1.4	34	3.4	
Jordan	40	3.6	⇧	28	1.3	37	2.0	⇧
Botswana	38	4.2	⇧	27	1.0	33	2.6	⇧
Romania	30	5.6		20	1.3	20	8.1	
Oman	26	2.2	⇧	15	1.0	32	3.4	⇧
Turkey	19	4.8		10	1.0	12	3.4	
Malta	32	3.2	⇧	23	0.8	29	2.6	⇧
Tunisia	29	3.8		21	1.6	27	3.6	
Bosnia and Herzegovina	32	1.9	⇧	24	1.2	36	3.9	⇧
Sweden	61	3.1		55	1.7	57	2.6	
Egypt	20	1.5	⇧	13	0.8	20	3.2	⇧
El Salvador	25	4.0		18	1.3	29	4.6	⇧
Georgia	63	6.3		57	2.8	61	4.4	
Saudi Arabia	37	2.6	⇧	31	1.4	36	2.7	
Kuwait	48	2.3		44	1.6	39	2.6	
Thailand	20	7.9		16	1.4	13	3.6	
Lithuania	36	4.3		33	1.2	32	3.4	
Russian Federation	45	4.3		42	1.5	46	2.4	
Israel	52	3.8		49	1.8	56	2.4	⇧
Lebanon	38	2.8		35	1.9	50	4.2	⇧
Ghana	18	1.7		16	1.2	21	3.0	
Qatar	55	1.3		54	0.8	47	1.4	⇩
Chinese Taipei	36	4.5		35	2.0	36	5.1	
Syria, Arab Republic of	26	1.9		25	1.4	29	3.3	
Iran, Islamic Republic of	16	6.6		16	1.4	13	3.3	
Malaysia	20	2.5		21	1.6	20	3.1	

(continued)

Table 4B.7 (continued)

Country	First generation immigrant			Native			Second generation immigrant		
	Percent	SE	sig	Percent	SE		Percent	SE	sig
Hungary	31	5.1		32	1.4		47	4.6	⇧
Colombia	19	3.2		21	1.2		28	5.3	
Indonesia	9	1.2	⇩	14	1.2		16	6.6	
Ukraine	63	3.6		68	1.4		75	2.4	⇧
Serbia	31	3.3		38	1.8		44	2.5	
Bulgaria	35	3.6		42	1.8		58	7.2	⇧
Slovenia	39	4.3		47	1.3		40	2.8	⇩
Norway	78	3.5		85	1.0		85	2.6	
Hong Kong, SAR	16	1.8	⇩	26	1.6		10	0.9	⇩
United States (Massachusetts)	60	3.1	⇩	74	1.9		59	3.0	⇩
United States	46	2.6	⇩	62	1.4		43	2.5	⇩
United States (Minnesota)	52	6.0	⇩	67	2.4		50	3.9	⇩
Mongolia	26	2.1	⇩	42	1.9		33	2.8	⇩
Armenia	38	3.9	⇩	58	1.8		61	4.4	

students. One might also suspect that the differences in the educational background are somewhat caused by the immigration policies of the different countries. I will investigate this in more detail in the country research in Chaps. 5A and 5B.

On the other side of the spectrum, I find in the United States, Hong Kong, Indonesia, Mongolia, and Armenia, where there are statistically significantly fewer students among first-generation immigrant students than among native students with at least one of the parents having achieved an ISCED level 5 education or higher. As we have seen in Table 4A.3, the mathematics achievement trend of first-generation immigrant students compared to the natives in Hong Kong is rather negative. It would be interesting to examine if there are changes in the educational background of the parents of immigrant students.

One of the important outcomes of this analysis is that in all but seven countries, the percentage of immigrant students with at least one of their parents having completed an ISCED level 5 education or higher is statistically no different from native students or even statistically significantly higher. Consequently, I cannot support the thesis that a lower educational background of the immigrant students is a factor that influences the – in general – lower achievement of first-generation immigrant students.

Comparing the situations of second-generation immigrant students and native students, I find eight countries with a statistically significantly higher percentage of native students with at least one of their parents having completed an education of ISCED level 5 or higher. The list of concerned countries is very similar to the one of first-generation immigrant students but includes also Korea, Slovenia, and Qatar – keeping in mind that the statistic for Korea is not reliable due to the small number of cases.

On the other hand, there are 16 countries where I find a statistically significantly higher percentage of second-generation immigrant students than native students with one of the parents having completed an ISCED level 5 education. This means that it is also not true that there are more students among second-generation immigrant students whose parents are educated less well.

After examining the percentage of the students with at least one of the parents having finished an ISCED level 5 education, I now want to look at the percentage of students where both parents have a rather low educational level. Table 4B.8 shows the percentage of native students and first- and second-generation immigrant students, neither of whose parents completed an education above ISCED level 2. As stated above, an ISCED level 2 education can be regarded as a very minimal education that corresponds to the compulsory education in many countries. As for Table 4B.7, the countries are indicated that show statistically significant differences for immigrant students compared to native students.

From Table 4B.8 I can derive that in 13 countries, there are statistically significantly more students among first-generation immigrants than native students, both of whose parents haven't completed any education beyond ISCED level 2. The list of concerned countries is of course quite similar to the abovementioned list of countries with a statistically significant higher percentage of native students with at least one of the parents having completed an ISCED level 5 education or higher. Again, the list includes participants such as the United States, Hong Kong, Mongolia, and Armenia but also Hungary, Slovenia, Norway, Chinese Taipei, Serbia, and Bahrain.

But then there are seven countries where there are statistically significantly more native students compared to first-generation immigrant students whose parents have not completed any education beyond ISCED level 2. The group of countries concerned is quite diverse, including Dubai, Turkey, Botswana, Oman, Palestine, Singapore, and Cyprus. As for the countries with statistically significantly more students among first-generation immigrants than natives with at least one of the parents having completed an education of ISCED level 5 or higher, we find Singapore and Dubai on the list – two countries with relatively good achievement results of immigrant students.

I also want to discuss the results for second-generation immigrant students. Here, I find ten countries with a statistically significantly higher percentage of native students than second-generation immigrant students, neither of whose parents have completed an education of ISCED level 2.

On the other hand, there are 13 countries for which I find a statistically significantly higher percentage of second-generation immigrant students than native students with neither parent having completed an ISCED level 2 education.

Compared to the statistics for second-generation immigrant students related to the parents with a high educational background, this result seems to be very ambiguous. Also in terms of the percentage of parents with a very low educational level, I cannot conclude that there is a general tendency that more second-generation immigrants than native students are affected.

Table 4B.8 Percentage of students whose parents' highest education is ISCED level 2 and lower

Country	First generation immigrant			Native		Second generation immigrant		
	Percent	SE	sig	Percent	SE	Percent	SE	sig
United Arab Emirates (Dubai)	5	0.7	⇩	21	1.7	13	1.6	⇩
Turkey	53	7.1	⇩	69	1.8	60	5.7	
Botswana	31	4.0	⇩	41	1.1	31	2.8	⇩
Oman	49	2.4	⇩	58	1.6	43	4.1	⇩
El Salvador	51	5.1		58	1.8	43	5.0	⇩
Palestinian National Authority	16	1.4	⇩	23	1.2	19	2.5	
Singapore	10	1.2	⇩	15	0.9	19	1.3	⇧
Iran, Islamic Republic of	55	10.0		60	1.9	70	5.3	
Cyprus	10	1.5	⇩	15	0.7	11	1.3	⇩
Australia	18	2.9		22	1.4	19	1.7	
Korea, Republic of	-	-		4	0.4	-	-	
Morocco	54	7.1		58	2.1	49	4.9	
Tunisia	38	3.8		41	1.7	43	4.3	
Italy	27	3.6		30	1.4	24	3.2	
Qatar	20	1.0		22	0.6	25	1.0	⇧
Bosnia and Herzegovina	12	1.6		14	1.1	11	2.8	
Saudi Arabia	41	3.1		43	1.8	37	3.1	
Malta	50	3.1		52	0.9	44	3.1	⇩
Canada (Quebec)	4	1.4		5	0.5	6	1.2	
Canada (British Columbia)	3	1.0		4	0.6	4	0.8	
Syria, Arab Republic of	36	2.0		37	1.6	31	2.7	
Canada (Ontario)	3	1.3		3	0.7	4	0.9	
Jordan	20	1.8		20	1.3	15	1.7	⇩
Lebanon	37	3.3		37	2.4	24	4.9	⇩
Malaysia	30	2.6		28	1.7	37	3.9	⇧
Kuwait	15	1.9		14	1.0	20	2.0	⇧
Lithuania	7	2.1		6	0.6	6	2.0	
Ukraine	7	1.9		6	0.5	4	1.0	
Egypt	49	1.9		47	1.6	48	4.9	
Russian Federation	8	2.2		6	0.6	6	1.7	
Israel	15	2.4		12	1.0	13	1.6	
Georgia	5	1.8		3	0.4	3	1.4	
Romania	18	6.2		13	1.3	32	12.0	
Bahrain	27	2.0	⇧	22	1.0	39	3.0	⇧
Sweden	13	2.3		10	0.9	14	1.8	⇧
Ghana	45	3.0		41	1.5	36	3.9	
Indonesia	60	2.7		55	2.0	67	8.4	
Thailand	68	8.9		62	1.6	68	6.1	

(continued)

Table 4B.8 (continued)

Country	First generation immigrant			Native		Second generation immigrant		
	Percent	SE	sig	Percent	SE	Percent	SE	sig
Bulgaria	15	3.2		9	1.4	9	4.2	
Czech Republic	9	3.4	⇧	2	0.3	2	1.0	
Serbia	14	3.3	⇧	7	1.0	7	1.3	
Japan	9	5.6		2	0.3			⇩
Chinese Taipei	26	3.6	⇧	18	1.4	27	5.0	
Norway	11	3.0	⇧	3	0.5	8	1.9	⇧
Slovenia	13	3.1	⇧	5	0.5	9	1.4	⇧
Mongolia	26	2.2	⇧	17	1.1	22	2.1	⇧
Colombia	57	4.4		48	1.6	46	4.5	
United States (Massachusetts)	12	1.9	⇧	3	0.5	12	2.9	⇧
United States (Minnesota)	17	5.1	⇧	3	0.7	15	2.7	⇧
Armenia	16	3.7	⇧	1	0.3	4	1.3	⇧
Hungary	24	6.2	⇧	8	0.8	3	1.3	⇩
United States	26	2.3	⇧	6	0.5	28	2.3	⇧
Hong Kong, SAR	45	1.8	⇧	25	1.4	46	2.0	⇧

For this group of students, neither of whose parents have completed more than an ISCED level 2 education, the picture is not that clear. In 14 countries a significantly higher percentage of second-generation immigrant students compared with native students have parents with an education of ISCED level 2 or lower. The opposite is true in 12 countries.

There is a significantly higher percentage of first-generation immigrants than native students whose parents completed only an education of ISCED level 2 or lower, whereas the opposite is true in only eight countries.

I conclude that the parents of neither first- nor second-generation immigrant students are less educated than the parents of native students. Therefore I cannot support the thesis pursued by the OECD (2010a, 2010c) that immigrant students are performing lower than native students because of lower SES background – at least not with respect to the parents' education. There is rather a strong tendency that first-generation immigrant students have parents with a very high level of education. In 19 countries, first-generation immigrant students have a significantly higher percentage of at least one of their parents having completed an education of ISCED level 5 or higher than native students. And in only seven countries, I find a significantly higher percentage of native students with at least one of their parents having completed an ISCED level 5 education or higher compared to first-generation immigrant students. One might suspect that immigration policies in some countries favor immigrants with a higher education which results in a higher percentage of immigrants with high educational levels. This will be investigated further in Chaps. 5A and 5B.

After examining the parents' education as one aspect of the socioeconomic background, I want to focus on other SES aspects. As explained in the review in Chap. 2, another aspect of the SES background that was found to discriminate quite well between students with higher SES background and lower SES background is the number of books at home (see Postlethwaite and Ross (1992)). This will be my next focus.

Home Possessions

The students in TIMSS were asked about the number of books at their home with five response options (Fig. 4B.4):

For the following analysis, this ordinal scale was transferred into an interval scale by recoding all categories to the average number of books that is covered in each category. The responses to the last category were recoded to the minimum amount plus the difference to the mean of the second last category following the traditional approach as used, for example, by Elley (1994). This results in the following recodings: 1->5, 2-> 18, 3->63, 4->150, 5->250.

Table 4B.9 shows the average number of books at home for native students and for first- and second-generation immigrant students. It is also indicated if the average for one of the immigrant student groups is statistically significantly higher or lower than the average for native students. As can be seen in Table 4B.9, there are 23 countries where the number of books for first-generation immigrant students is statistically significantly lower than for native students. In five countries the average is significantly higher for first-generation immigrant students than for native students.

About how many books are there in your home? (Do not count magazines, newspapers, or your school books.)

*Fill in **one** circle only*

None or very few
(0-10 books) ------------------------ ①

Enough to fill one shelf
(11-25 books) ----------------------- ②

Enough to fill one bookcase
(26-100 books) ---------------------- ③

Enough to fill two bookcase
(101-200 books) --------------------- ④

Enough to fill three or more bookcases
(more than 200 books) ---------------- ⑤

Fig. 4B.4 TIMSS 2007 student questionnaire, variable AS4GBOOK (Source: TIMSS 2007 Assessment. Copyright © 2009 International Association for the Evaluation of Educational Achievement (IEA). Publisher: TIMSS & PIRLS International Study Center, Lynch School of Education, Boston College)

Table 4B.9 Number of books at home by the students' immigration status

Country	First generation immigrant			Native students		Second generation immigrant		
	Mean	SE		Mean	SE	Mean	SE	
Korea, Republic of	191	24.8	⇑	123	2.1	58	19.9	⇓
Morocco	68	12.1		46	2.6	60	9.0	
Iran, Islamic Republic of	58	18.6		40	1.9	30	5.3	
Botswana	53	5.3	⇑	37	1.3	46	4.0	⇑
United Arab Emirates (Dubai)	81	3.1	⇑	65	3.3	66	3.6	
Thailand	52	14.9		36	1.8	31	6.2	
Tunisia	49	5.2	⇑	37	1.4	41	4.3	
Turkey	57	8.1		48	1.9	56	7.3	
Ghana	44	2.4	⇑	37	1.6	39	6.3	
Jordan	66	5.7		59	2.0	70	2.9	⇑
El Salvador	37	5.4		32	1.6	49	7.4	⇑
Palestinian National Authority	53	2.7		49	2.1	45	3.5	
Kuwait	63	3.6		59	1.5	60	3.5	
Singapore	83	4.1		80	1.8	87	3.4	
Bosnia and Herzegovina	39	2.5		36	1.1	45	4.0	⇑
Qatar	85	2.1		82	1.7	84	2.2	
Syria, Arab Republic of	47	2.5		45	1.4	46	4.0	
Canada (Quebec)	74	5.5		72	2.3	84	6.4	
Saudi Arabia	56	4.5		55	2.0	61	3.6	
Lithuania	72	5.6		71	1.8	71	5.2	
Ukraine	80	5.2		79	2.2	99	4.2	⇑
Georgia	96	13.8		96	3.9	87	10.1	
Egypt	41	2.0		42	1.4	41	5.4	
Scotland	77	9.0		78	2.6	97	6.6	⇑
Lebanon	63	4.8		64	2.4	77	7.7	
Indonesia	26	1.4		29	1.0	37	8.1	
Mongolia	27	2.7		31	1.5	25	2.1	⇓
Bahrain	71	3.5		75	1.3	61	3.5	⇓
Colombia	31	3.9		36	1.6	39	8.0	
Malta	98	4.7		102	1.3	99	3.6	
Canada (British Columbia)	112	4.4		120	2.8	106	5.1	⇓
Oman	56	3.0	⇓	64	2.2	70	4.4	
Malaysia	45	3.9	⇓	54	2.2	48	5.2	
Russian Federation	88	5.9		98	2.0	100	5.3	
Romania	51	6.8		64	2.3	55	11.3	
Serbia	45	4.3	⇓	59	1.9	60	2.9	
Slovenia	69	5.4	⇓	84	1.7	67	3.5	⇓
Canada (Ontario)	98	5.8	⇓	116	4.3	112	4.3	

(continued)

Table 4B.9 (continued)

Country	First generation immigrant			Native students		Second generation immigrant		
	Mean	SE		Mean	SE	Mean	SE	
Cyprus	68	4.7	⇩	87	1.6	84	3.7	
England	76	6.3	⇩	96	2.7	93	5.0	
Italy	77	6.2	⇩	103	2.7	112	6.0	
Australia	90	5.7	⇩	116	2.8	111	4.6	
Chinese Taipei	64	7.2	⇩	91	3.2	81	9.3	
Israel	82	4.9	⇩	109	2.6	108	4.4	
Japan	61	11.1	⇩	90	1.9	83	12.5	
Bulgaria	73	6.8	⇩	102	2.8	109	14.4	
Norway	90	6.5	⇩	119	2.3	103	5.6	⇩
Armenia	64	6.0	⇩	94	2.4	89	5.9	
Hong Kong, SAR	46	2.4	⇩	78	2.5	53	2.0	⇩
Czech Republic	61	8.6	⇩	93	1.6	83	4.5	⇩
United States (Massachusetts)	88	10.5	⇩	125	3.8	90	6.7	⇩
Hungary	83	11.2	⇩	120	2.5	121	8.4	
United States	60	3.2	⇩	103	2.5	68	3.0	⇩
Sweden	81	5.1	⇩	126	2.5	103	4.0	⇩
Spain (Basque Country)	79	9.0	⇩	125	2.7	119	8.6	
United States (Minnesota)	69	9.7	⇩	120	3.8	73	4.9	⇩

For second-generation immigrant students, the results are not that unambiguous. Only in 12 countries, the number of books at home is statistically significantly lower than for native students, whereas it is statistically significantly higher in six countries.

If we accept the number of books at home as an indicator for the socioeconomic capital of the students' family – although probably a weak one – we can conclude that nearly half of the countries' first-generation immigrant students come from a less affluent background than native students. But as already cautioned by Hansen and Gustafsson (2010), the number of books at home might, as an indicator, work differently for different immigration groups and comparisons should be made with great care.

But since we know that the socioeconomic status predicts student achievement quite well, it is even more interesting to investigate if and how the socioeconomic status – measured by the number of books at home – relates to the achievement for the different groups of students.

To investigate this, a regression analysis was conducted. For native students and first- and second-generation immigrant students, the mathematics achievement regressed on the number of books at home. For the number of books at home, the recoded variable was used.

Table 4B.10 shows the beta-coefficients of the regression analysis. The regression coefficients are highlighted if they are statistically significantly greater than zero. It is also marked if the regression coefficients for one of the immigrant student groups are statistically significantly different from the regression coefficients for native students. Since the recoded variable is used, one can interpret the regression coefficients in the following way: a value of one indicates that students with one or more books at home achieved one score more point on the TIMSS 2007 mathematics scale.

The results show that for native students in all countries, the regression coefficients are statistically significantly positive. This is not surprising and in line with what is already reported in the TIMSS 2007 international mathematics report (see exhibit 4.4 in (Mullis et al., 2008)).

The results for immigrant students are mostly similar. In most of the countries, the regression coefficients are statistically significantly positive. In most other cases, the regression coefficients are also positive but not statistically significant. This is not too surprising since the standard errors are larger for the immigrant populations due to the smaller sample sizes in these countries. There are also three countries, Armenia, Ghana, and Kuwait, where the regression coefficients are negative for first-generation immigrant students but not statistically different from zero.

When comparing the regression coefficients for immigrant students to the ones for native students, I find six countries where the regression coefficients for first-generation immigrant students are statistically significantly lower than for native students and three countries where they are statistically significantly higher. The countries where they are higher are Chinese Taipei, Malta, and Norway. In these countries, the SES background of the students measured by the number of books at home is more strongly related to the mathematics achievement among first-generation immigrant students than among native students.

For second-generation immigrant students, we find three participants where the regression coefficients for second-generation immigrant students are statistically significantly lower than for native students, and two countries where they are statistically significantly higher. The participants with the lower regression coefficients for immigrant students are Tunisia, the Canadian state of British Columbia, and the US state of Minnesota. The case of British Columbia will be discussed further in Chap. 5B – the results match with other researches on the influence of SES background on parents motivating their students and resulting higher achievement in British Columbia. The result for Minnesota is surprising in the sense that it contradicts the results for Massachusetts where exactly the opposite can be observed since Massachusetts is one of the two countries where the regression coefficients for second-generation immigrant students are statistically significantly higher than for native students. The other one is Romania where, as discussed already earlier, the results are more of an artifact created by a very small sample size.

Overall I conclude that the number of books at home as a predictor of the SES background of the students works quite similar for native students as well as for first- and second-generation immigrant students. Students with more books at home achieve better in TIMSS mathematics than students with fewer books at home. The achievement difference for students with a different number of books at home varies between immigrant students and native students in some countries, but the general

Table 4B.10 Regression coefficients of number of books at home on mathematics achievement for native and immigrant students

Country	First generation immigrant			Native		Second generation immigrant		
	Beta	SE	sign	Beta	SE	Beta	SE	sign
Thailand	0.38	0.29		0.58	0.07	0.19	0.29	
Turkey	0.59	0.29		0.52	0.04	0.58	0.20	
Romania	0.52	0.13		0.47	0.03	1.21	0.32	⇧
Colombia	0.32	0.19		0.41	0.04	0.66	0.22	
England	0.31	0.08		0.40	0.03	0.35	0.05	
Korea, Republic of	0.25	0.31		0.40	0.02	0.22	0.36	
Malaysia	0.30	0.08		0.38	0.03	0.47	0.13	
Chinese Taipei	0.63	0.10	⇧	0.38	0.02	0.26	0.13	
Scotland	0.48	0.07		0.37	0.03	0.35	0.06	
Hungary	0.58	0.12		0.37	0.02	0.51	0.09	
Tunisia	0.22	0.08		0.37	0.03	0.14	0.11	⇩
Singapore	0.31	0.06		0.36	0.03	0.35	0.05	
Lithuania	0.26	0.09		0.34	0.02	0.34	0.06	
Iran, Islamic Republic of	0.30	0.17		0.34	0.04	0.16	0.24	
Serbia	0.26	0.14		0.33	0.03	0.43	0.05	
Indonesia	0.14	0.14		0.32	0.07	0.95	0.66	
Bulgaria	0.17	0.07	⇩	0.32	0.04	0.42	0.17	
Czech Republic	0.43	0.11		0.31	0.02	0.43	0.07	
Australia	0.40	0.05		0.30	0.02	0.31	0.04	
Ukraine	0.38	0.08		0.29	0.03	0.29	0.05	
Malta	0.47	0.07	⇧	0.29	0.02	0.23	0.05	
United States (Massachusetts)	0.36	0.07		0.29	0.03	0.42	0.05	⇧
Bosnia and Herzegovina	0.30	0.05		0.29	0.04	0.37	0.12	
United States	0.32	0.06		0.28	0.01	0.33	0.03	
Mongolia	0.27	0.11		0.27	0.03	0.25	0.07	
Canada (Quebec)	0.29	0.08		0.27	0.03	0.19	0.06	
Hong Kong, SAR	0.27	0.05		0.27	0.04	0.23	0.05	
United Arab Emirates (Dubai)	0.34	0.03		0.27	0.05	0.22	0.04	
Slovenia	0.30	0.05		0.26	0.02	0.27	0.06	
Japan	0.67	0.26		0.26	0.02	0.34	0.14	
Georgia	0.35	0.11		0.25	0.03	0.34	0.12	
Russian Federation	0.10	0.06	⇩	0.25	0.03	0.36	0.05	
Jordan	0.30	0.09		0.25	0.03	0.23	0.05	
Israel	0.19	0.06		0.25	0.03	0.22	0.05	
Cyprus	0.35	0.06		0.25	0.02	0.28	0.06	
Canada (Ontario)	0.24	0.05		0.24	0.04	0.21	0.03	
Canada (British Columbia)	0.18	0.05		0.24	0.02	0.16	0.03	⇩

(continued)

Table 4B.10 (continued)

Country	First generation immigrant			Native		Second generation immigrant		
	Beta	SE	sign	Beta	SE	Beta	SE	sign
Spain (Basque Country)	0.32	0.08		0.23	0.02	0.32	0.06	
Sweden	0.32	0.05		0.23	0.02	0.19	0.03	
United States (Minnesota)	0.38	0.10		0.22	0.03	0.09	0.05	⇩
El Salvador	0.21	0.15		0.22	0.05	0.14	0.13	
Italy	0.13	0.07		0.22	0.02	0.20	0.06	
Oman	0.15	0.06		0.21	0.03	0.26	0.05	
Norway	0.30	0.04	⇧	0.20	0.02	0.22	0.03	
Morocco	0.10	0.10		0.20	0.03	0.20	0.10	
Palestinian National Authority	0.08	0.06		0.17	0.04	0.32	0.13	
Saudi Arabia	0.10	0.06		0.17	0.02	0.13	0.06	
Bahrain	0.21	0.06		0.16	0.03	0.18	0.05	
Qatar	0.11	0.03		0.16	0.02	0.13	0.03	
Lebanon	0.12	0.07		0.16	0.04	0.11	0.10	
Egypt	0.02	0.05		0.15	0.04	0.04	0.13	
Armenia	−0.10	0.09	⇩	0.12	0.02	0.21	0.10	
Syria, Arab Republic of	0.05	0.05		0.12	0.04	0.06	0.10	
Botswana	0.23	0.14		0.11	0.03	0.06	0.07	
Ghana	−0.03	0.05	⇩	0.10	0.04	0.19	0.14	
Kuwait	−0.02	0.05	⇩	0.09	0.03	0.06	0.07	

tendency is the same. This is somewhat surprising as we know that the population of immigrant students is culturally quite diverse and using the number of books at home as an SES predictor does not necessarily work across cultures. A more in-depth analysis of the number of books at home and the varying achievement results for the three groups of students was not possible because the small number of observations for the immigrant populations and the number of response categories for the books at home resulted in quite unstable estimates.

Summarizing the results of the analysis of the students' socioeconomic background, I conclude that there are no differences between immigrant students and native students regarding parental education. Quite the contrary, I even found a couple of countries where the parental education was statistically significantly better for first-generation immigrant students than for native students. The results for the number of books at home are clearly different. I find that in a couple of countries, first-generation immigrant students have a significantly smaller number of books at home than native students. I also find that the number of books at home is a good predictor of mathematics achievement for native students as well as for first- and second-generation immigrant students.

Students' Attitudes

As described in the corresponding part of Chap. 2, students' attitudes are considered as influencing as well as being influenced by the students' achievement. But as elaborated in Chap. 2, positive attitudes are also considered as a positive outcome of the students in themselves. The next analysis will investigate the attitudes towards the school, the subject mathematics, and the students' self-efficacy in mathematics. My aim is to find out if there are differences in these attitudinal scales between native students and first- and second-generation immigrant students.

Students' Attitudes to School

The first focus is on the students' attitude towards the school in general. In TIMSS 2007, grade eight students were asked the following question (Fig. 4B.5):

Table 4B.11 shows the percentages of students who answered that they agree a lot or agree a little to this question for each of the immigrant groups together with an indicator whether the percentage differs statistically significantly between immigrant and native students.

The table indicates that a higher percentage of first-generation immigrants than native students like being in school in the Canadian provinces, Qatar, Spain (Basque Country), Sweden, and Dubai. For Armenia, Botswana, Georgia, Ghana, Hong Kong, Jordan, Lebanon, Malaysia, Oman, Syria, and Tunisia, the percentage is smaller for first-generation immigrants than for native students. So overall, there are more countries where the attitudes towards school among first-generation immigrant students are less positive than among native students.

For second-generation immigrant students, this is quite different. TIMSS participants where the percentage of students with positive attitudes towards the school is statistically significantly higher for second-generation immigrant students than for native students are Australia, British Colombia, Quebec, England, Korea, Mongolia,

How much do you agree with these statements about your school?

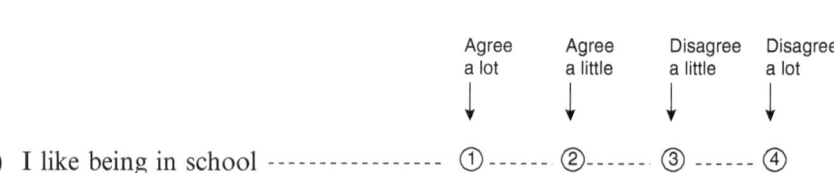

*Fill in **one** circle for each line*

| | Agree a lot | Agree a little | Disagree a little | Disagree a lot |

a) I like being in school -------------- ① ------ ② ------ ③ ------ ④

Fig. 4B.5 TIMSS 2007 student questionnaire, variable BS4GALBS (Source: TIMSS 2007 Assessment. Copyright © 2009 International Association for the Evaluation of Educational Achievement (IEA). Publisher: TIMSS & PIRLS International Study Center, Lynch School of Education, Boston College)

Table 4B.11 Percentage of students who like to go to school by immigration background

Country	First generation immigrant			Native		Second generation immigrant		
	perc.	SE	sign	perc.	SE	perc.	SE	sign
Spain (Basque Country)	72	4.4	⇧	50	1.9	40	5.3	
United Arab Emirates (Dubai)	87	1.1	⇧	73	2.0	83	1.6	⇧
Canada (British Columbia)	83	1.6	⇧	71	1.4	82	1.3	⇧
United States (Minnesota)	88	4.0	⇧	76	1.7	80	2.6	
Canada (Quebec)	73	2.5	⇧	63	1.5	76	2.4	⇧
Canada (Ontario)	81	2.5	⇧	72	1.8	75	1.6	
Czech Republic	63	4.9		56	1.2	58	3.7	
Qatar	78	1.0	⇧	72	0.8	79	1.0	⇧
Sweden	72	2.7	⇧	66	1.1	71	1.9	⇧
Korea, Republic of	76	8.9		71	0.9	100	0.0	⇧
Ukraine	91	2.0		87	0.9	81	1.4	⇩
England	76	2.8		73	1.3	79	1.8	⇧
Cyprus	65	2.3		62	1.0	62	2.5	
Australia	75	2.2		73	1.2	77	1.7	⇧
United States	74	1.8		72	0.9	75	1.4	⇧
Bahrain	78	2.1		76	0.9	80	1.9	
Italy	66	3.9		65	1.2	62	3.2	
Lithuania	70	3.7		68	1.1	67	3.4	
Bulgaria	75	3.6		74	1.1	64	6.1	
Russian Federation	81	2.6		80	1.3	78	2.4	
Iran, Islamic Republic of	92	4.6		92	0.6	80	5.7	⇩
Singapore	86	1.5		86	0.8	85	1.4	
Egypt	95	0.6		95	0.4	93	1.5	
Romania	87	3.2		87	0.7	76	10.3	
El Salvador	96	1.4		96	0.4	96	1.7	
Saudi Arabia	82	2.4		82	1.3	86	1.9	
Mongolia	92	1.0		93	0.6	96	0.9	⇧
Indonesia	97	0.7		98	0.3	95	3.1	
Norway	74	2.6		75	1.1	77	1.9	
Colombia	94	3.2		96	0.4	95	2.2	
Bosnia and Herzegovina	76	1.8		78	1.2	78	3.5	
Israel	71	2.3		74	1.4	70	2.2	
Serbia	63	3.7		65	1.8	62	2.3	
Morocco	95	1.7		98	0.5	98	1.0	
Kuwait	73	2.2		75	1.0	77	1.9	
United States (Massachusetts)	67	4.2		70	1.9	72	2.7	
Malta	57	2.7		60	0.8	59	2.1	
Syria, Arab Republic of	94	0.8	⇩	97	0.3	95	1.4	
Scotland	65	3.6		69	0.8	75	3.1	⇧
Palestinian National Authority	86	1.7		90	1.1	90	2.1	

(continued)

Table 4B.11 (continued)

Country	First generation immigrant			Native		Second generation immigrant		
	perc.	SE	sign	perc.	SE	perc.	SE	sign
Oman	90	1.4	⇩	94	0.5	95	1.2	
Japan	71	7.7		76	0.9	71	7.6	
Slovenia	46	4.6		50	1.2	54	2.5	
Thailand	86	7.0		91	0.5	86	3.4	
Lebanon	82	1.6	⇩	87	1.1	82	3.0	
Ghana	93	1.1	⇩	99	0.3	96	1.7	
Jordan	86	1.8	⇩	92	0.9	88	1.3	⇩
Georgia	89	2.5	⇩	95	0.5	92	3.4	
Chinese Taipei	60	3.2		66	1.1	64	6.3	
Armenia	81	3.0	⇩	88	0.8	79	3.1	⇩
Malaysia	82	2.5	⇩	89	0.7	92	1.8	
Turkey	89	5.3		96	0.3	93	2.6	
Tunisia	84	3.5	⇩	91	0.6	86	2.9	
Hong Kong, SAR	67	2.0	⇩	75	1.5	73	1.6	
Hungary	58	6.9		66	1.3	71	6.2	
Botswana	83	2.6	⇩	96	0.3	88	1.5	⇩

Qatar, Scotland, Sweden, Dubai, and the United States as opposed to Armenia, Botswana, Iran, Jordan, and Ukraine where the opposite is true. Interestingly the attitudes towards school are more positive among second-generation immigrants than native and first-generation immigrants.

Especially interesting cases are those of British Columbia, Quebec, and Dubai where for both immigrant groups, the percentage of students with positive attitudes towards school exceeds the percentage for native students by more than 10 %. Another very interesting case is that of the Basque region of Spain where 40 % of second-generation immigrant students, 50 % of native students, and 72 % of the first-generation immigrant students have positive attitudes towards school.

For the interpretation of these results, we need to be careful and must consider that the students come from different cultural backgrounds which might influence their response pattern. But overall we can determine that for the large majority of educational systems, there is no difference in terms of attitudes towards school between immigrant students and native students. However, I can observe some countries where there are significant differences – in some cases in favor of native students, in others of immigrant students. The results for some of the countries should be evaluated further to understand why these differences emerged.

Students' Attitude Towards Mathematics

Next I will examine the attitudes towards mathematics. The TIMSS student questionnaire includes a question on how much students agree with several statements about learning mathematics. One of them is if they enjoy learning mathematics, and another one is if they like mathematics (Fig. 4B.6).

Table 4B.12 shows the percentages of students with immigrant status who agree or strongly agree to "I enjoy learning mathematics." As it was the case with the statement "I like being in school" examined earlier, I cannot find a clear pattern here, either. The average for all three groups across all participating countries is 65 %. But again I can find countries where the results for immigrant students and native students differ clearly. Interestingly the pattern is very similar to the one about liking to go to school. Again, in the Canadian provinces, the students from both immigrant groups have more positive attitudes than native students, and this is also true for, again, Qatar, Spain (Basque Country), Sweden, and Dubai. Moreover, there are more countries with statistically significantly lower percentages of students who agree or strongly agree to "I enjoy learning mathematics" in the first-generation immigrant group compared to the native students. For second-generation immigrants, the number with statistically significantly higher percentages of students who like mathematics is the same as the number with statistically significantly lower percentages compared to native students. The set of countries with higher percentages is again very similar for first- and second-generation immigrant students.

in b, eir.

Overall I conclude that native and immigrant students' attitudes towards mathematics are the same. I note that there are countries where the attitudes of immigrant

How much do you agree with these statements about learning mathematics?

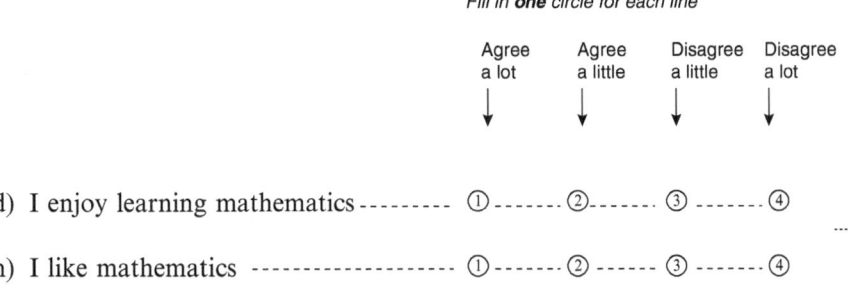

Fig. 4B.6 TIMSS 2007 student questionnaire, variables BS4MAENJ and BS4MALIK (Source: TIMSS 2007 Assessment. Copyright © 2009 International Association for the Evaluation of Educational Achievement (IEA). Publisher: TIMSS & PIRLS International Study Center, Lynch School of Education, Boston College)

Table 4B.12 Percentages of students who strongly agree or agree to enjoy learning mathematics

Country	First generation immigrant			Native		Second generation immigrant		
	Perc.	SE		Perc.	SE	Perc.	SE	
United Arab Emirates (Dubai)	76	1.4	⇧	61	2.0	75	1.8	⇧
Canada (Ontario)	75	2.7	⇧	62	2.1	68	2.0	⇧
Canada (British Columbia)	63	2.6	⇧	52	1.6	57	1.6	⇧
Spain (Basque Country)	59	4.5	⇧	48	1.7	43	5.3	
England	68	3.4	⇧	58	1.6	67	2.8	⇧
Australia	63	2.8	⇧	53	1.3	56	2.2	
Czech Republic	45	5.5		37	1.1	37	3.4	
Sweden	69	2.8	⇧	61	1.2	68	2.0	⇧
United States (Minnesota)	68	4.7		61	2.6	69	3.4	
Singapore	82	2.0	⇧	75	0.9	72	1.8	
United States	65	1.6	⇧	59	1.0	60	1.5	
Qatar	76	1.1	⇧	70	0.8	77	0.9	⇧
Norway	65	3.6		60	1.0	65	2.0	⇧
Saudi Arabia	78	1.7	⇧	73	1.5	75	2.2	
Canada (Quebec)	62	3.4		57	1.9	65	2.8	⇧
Ukraine	64	3.7		60	1.6	57	2.3	
Kuwait	76	1.9		73	1.1	79	1.7	⇧
Bahrain	80	2.0		78	0.9	76	2.1	
Scotland	59	3.0		56	1.2	60	3.6	
Russian Federation	59	4.2		57	1.2	54	2.7	
Slovenia	30	3.3		29	1.2	32	2.0	
Morocco	94	1.4		94	0.5	96	1.9	
Italy	57	3.5		58	1.2	53	3.5	
Hong Kong, SAR	61	2.3		62	1.6	60	1.5	
Colombia	89	3.1		90	0.7	82	3.7	⇩
Japan	39	9.1		40	1.2	39	5.9	
Indonesia	86	1.5		87	0.8	85	5.5	
Serbia	31	4.4		33	1.4	30	1.8	
Korea, Republic of	37	10.5		39	0.9	9	10.0	⇩
Turkey	85	4.6		87	0.9	89	3.4	
Jordan	84	2.0		86	1.1	84	1.7	
Hungary	40	5.5		42	1.5	45	5.4	
Oman	88	1.4		90	0.6	89	1.8	
Bosnia and Herzegovina	36	2.1		39	1.2	43	4.2	
El Salvador	82	2.9		85	0.9	86	3.3	
Ghana	84	1.6	⇩	87	0.9	79	2.9	⇩
Egypt	88	1.2	⇩	92	0.6	88	3.2	
Romania	53	4.2		58	1.5	55	11.1	
United States (Massachusetts)	54	4.1		59	2.2	67	3.4	⇧

(continued)

Table 4B.12 (continued)

Country	First generation immigrant			Native		Second generation immigrant		
	Perc.	SE		Perc.	SE	Perc.	SE	
Palestinian National Authority	70	1.9	⇩	75	1.4	73	2.7	
Lebanon	72	2.5	⇩	78	1.1	78	3.7	
Lithuania	47	4.9		53	1.3	55	3.8	
Mongolia	80	1.8	⇩	86	1.0	79	2.5	⇩
Bulgaria	53	5.4		60	1.4	42	5.8	⇩
Malaysia	72	2.2	⇩	79	1.0	80	3.3	
Syria, Arab Republic of	80	1.4	⇩	87	0.8	79	3.1	⇩
Israel	56	3.6	⇩	64	1.3	60	2.5	
Malta	49	2.9	⇩	58	0.9	52	2.1	⇩
Georgia	61	5.9		69	1.5	70	4.3	
Tunisia	80	3.1	⇩	89	0.6	88	2.5	
Cyprus	45	2.7	⇩	54	1.1	47	2.5	⇩
Chinese Taipei	35	3.1	⇩	45	1.3	50	6.3	
Armenia	54	2.6	⇩	66	1.4	64	3.8	
Thailand	67	8.2		81	0.9	76	4.1	
Botswana	70	3.7	⇩	86	0.8	75	2.3	⇩
Iran, Islamic Republic of	52	11.0	⇩	82	0.9	65	5.2	⇩

students towards mathematics are more positive, whereas in other countries the attitudes of native students are more positive. The Canadian provinces and Singapore are again among the participants where first-generation immigrant students have a more positive outcome – this time with respect to the students' attitudes.

If I now examine the percentages of students who indicated that they like mathematics by immigration status and also by sex, some interesting results emerge. Table 4B.13 shows the percentages of boys and girls by immigration status who agreed or strongly agreed that they like mathematics. Also included is an indicator for each of the three student groups that displays if the differences between boys and girls are significant within the group. Although the average of students who like mathematics across all countries is about the same for boys and girls and for all immigration statuses at about 64 % (with a slightly higher percentage for boys), some countries show very diverse results for the different immigration groups.

For example, in Bulgaria the percentage of native boys and girls is almost the same with 60 and 61 %, respectively. Forty-seven percent of first-generation immigrant girls and 59 % of first-generation immigrant boys like mathematics. For second-generation immigrants, the pattern is clearly opposite: whereas 57 % of the girls in this group reported they like mathematics, only 33 % of the boys did.

On the other hand, I find exactly the opposite pattern in Georgia where the girls' attitudes are more positive among first-generation immigrants and the boys' atti-

Table 4B.13 Percentage of boys and girls who like mathematics by immigration status

Country	First generation immigrant Boy Perc	SE		Girl Perc.	SE	Native Boy Perc.	SE		Girl Perc.	SE	Second generation immigrant Boy Perc.	SE		Girl Perc.	SE
Qatar	73	1.6		70	1.7	72	1.0	⇧	54	1.2	77	1.6	⇧	68	1.4
Chinese Taipei	37	4.4		29	5.0	52	1.6	⇧	36	1.5	51	7.8		37	8.0
England	67	5.1		63	5.7	62	1.7	⇧	50	2.0	67	3.7		58	3.7
Hong Kong, SAR	62	3.4	⇧	49	2.2	63	2.0	⇧	52	2.3	62	2.1	⇧	52	2.4
Japan	18	8.4	⇩	52	13.5	42	1.2	⇧	32	1.4	28	8.8		41	11.0
Kuwait	71	2.6		72	3.2	75	1.3	⇧	66	1.7	77	3.5		74	2.9
Korea, Republic of	62	10.0		30	14.2	46	1.2	⇧	38	1.4	-	-		-	-
Saudi Arabia	70	3.0		67	3.5	70	2.2	⇧	62	2.2	80	2.8	⇧	62	3.3
Italy	44	4.5	⇧	30	4.0	42	1.5	⇧	35	1.6	35	4.1		38	5.4
Palestinian National Authority	69	2.6		66	2.8	74	2.2	⇧	68	1.9	69	5.2		68	4.7
United Arab Emirates (Dubai)	74	2.1		70	3.1	58	4.1		52	3.0	76	3.9		67	2.5
Australia	57	3.5		55	4.2	53	2.0		47	2.5	57	2.8	⇧	48	3.0
El Salvador	82	4.0		75	5.3	86	1.0	⇧	81	1.2	78	5.7		80	6.2
Canada (Ontario)	79	3.0		73	3.8	66	2.2		61	2.6	69	3.0		63	2.6
Lebanon	77	2.8	⇧	68	3.1	81	1.6	⇧	77	1.4	80	4.5		73	6.1
Ghana	85	2.1		80	2.2	89	1.2	⇧	85	1.3	91	3.0	⇧	75	5.0
Egypt	87	1.2		87	1.6	92	0.8	⇧	88	1.1	91	3.0		88	3.8
Georgia	71	7.2		83	4.8	75	1.7		71	2.6	92	2.8	⇧	44	10.5
United States (Minnesota)	71	4.3		74	5.2	63	2.6		60	3.4	68	6.1		64	5.3
Tunisia	72	4.7		79	5.8	80	1.1		77	1.4	77	4.8		73	6.2
Morocco	88	3.9		88	3.6	89	0.9	⇧	86	1.1	86	4.3		84	4.1
Colombia	75	4.5		64	12.2	80	1.2		77	1.5	83	4.0	⇧	67	6.5
Indonesia	83	1.9		84	1.8	86	1.1		83	1.3	82	7.6		69	9.6
United States	62	2.6		61	2.7	61	1.0		58	1.3	59	2.5		54	2.0
Scotland	55	4.8		57	4.5	54	1.3		51	1.6	58	4.9		50	4.3
Syria, Arab Republic of	77	2.1		76	2.7	83	1.1		81	1.3	76	4.4		68	4.4
Canada (British Columbia)	63	2.9		57	3.5	51	2.0		48	1.8	60	2.4		56	2.3
Malta	51	4.2		53	4.3	54	1.1		52	1.1	54	3.0		50	2.6
Hungary	39	10.9		40	6.9	44	1.6		42	1.8	40	6.6		45	8.6
Spain (Basque Country)	60	5.6		55	6.1	55	1.9		54	2.2	48	7.8		50	7.1
Mongolia	82	2.5		83	2.2	86	1.1		85	1.2	76	3.5		79	3.0

(continued)

Table 4B.13 (continued)

| Country | First generation immigrant | | | | Native | | | | Second generation immigrant | | | |
| | Boy | | Girl | | Boy | | Girl | | Boy | | Girl | |
	Perc	SE	Perc.	SE	Perc.	SE	Perc.	SE	Perc.	SE	Perc.	SE
Canada (Quebec)	63	4.9	64	4.5	54	2.0	53	1.9	67	3.7	61	2.9
Sweden	61	4.5	49	4.3	44	1.7	43	1.5	52	3.1	46	2.9
United States (Massachusetts)	49	7.6	⇩ 69	4.0	62	2.4	61	2.1	69	4.0	65	3.4
Cyprus	52	3.7	52	3.5	63	1.1	63	1.3	54	3.6	55	3.9
Iran, Islamic Republic of	54	16.5	86	11.2	77	1.6	77	1.6	81	6.1	66	10.2
Bahrain	71	2.4	76	3.2	70	1.2	70	1.6	73	4.0	70	2.0
Slovenia	34	4.7	39	5.5	37	1.8	37	1.7	40	3.7	40	3.3
Norway	57	5.4	54	4.7	49	1.4	49	1.6	53	3.3	49	3.7
Botswana	82	4.5	80	4.5	86	1.2	87	0.9	78	2.7	82	3.2
Thailand	63	16.1	54	12.2	75	1.5	76	1.2	69	5.4	70	6.7
Bulgaria	59	4.9	47	6.6	60	1.5	61	1.8	33	8.2	⇩ 57	8.0
Jordan	77	3.5	77	3.7	82	2.2	83	1.6	84	1.8	84	2.5
Turkey	84	5.9	76	8.9	80	1.0	81	1.3	78	6.0	80	6.8
Singapore	81	2.2	80	2.7	71	1.3	73	1.4	71	2.5	69	2.8
Oman	85	1.4	89	2.0	88	1.2	90	0.8	85	2.8	91	2.1
Armenia	52	3.5	55	4.2	66	1.8	68	1.8	61	5.9	63	7.0
Bosnia and Herzegovina	46	2.6	51	2.5	51	1.8	54	1.9	43	6.1	52	6.3
Malaysia	77	3.9	74	4.1	77	1.0	80	1.1	75	4.9	79	4.3
Romania	44	7.2	⇩ 69	7.3	58	1.7	61	1.7	47	14.5	58	10.8
Israel	53	3.6	59	3.3	61	1.8	64	2.2	62	2.7	59	3.0
Czech Republic	39	6.7	57	7.7	41	1.7	⇩ 46	1.5	42	4.8	46	4.6
Lithuania	47	6.1	40	9.3	54	1.7	⇩ 59	1.7	58	5.7	58	3.5
Serbia	40	5.7	41	5.5	42	2.1	47	2.4	40	3.2	45	3.4
Ukraine	73	4.6	73	5.1	65	1.6	⇩ 71	1.9	64	2.6	66	3.2
Russian Federation	66	3.9	76	4.3	62	1.5	⇩ 70	1.5	65	3.8	59	4.0

tudes among second-generation immigrants. The same is true for Iran and the Russian Federation. More positive attitudes among girls in both immigration groups can be found in Japan and Romania. In Japan, one of the highest scoring countries in TIMSS, the positive attitudes towards mathematics are less prevalent than it is in general the case in high-achieving countries.[1] The general negative association between achievement and students' attitudes at country level is discussed, for example, in Shen and Tam (2008).

---

[1] However I must note that the statistics for Japan are not very reliable for the immigrant students because of the low number of immigrants in the sample.

I conclude that the students' attitudes towards learning and liking mathematics are in general similar between native students and immigrant students. Also between boys and girls, the percentages are similar for native and first- and second-generation immigrant students. At the individual country level, I do find countries where the attitudes towards mathematics are more positive for native students, immigrant students, immigrant boys, or native girls. Yet, considering students' attitudes towards mathematics as an outcome, I do not find any disadvantage for immigrant students – neither with respect to sex. Thus I cannot confirm the negative findings for girls' attitudes in previous research (see (Mata et al., 2012; Meece et al., 2006) for any of the student groups.

Students' Attitudes: Self-Efficacy

Previous research has shown that the students' self-rating regarding their abilities is in most cases quite accurate and matches the results from standardized tests. I discussed the importance of students' self-efficacy and its bidirectional connection to students' achievement in Chap. 4B.

The TIMSS student questionnaire has a question about the students' self-rating of their mathematics abilities (Fig. 4B.7):

As for the other students' attitude scales, the percentages of students agreeing or strongly agreeing are tabulated.

Table 4B.14 shows the percentage of students agreeing or strongly agreeing to usually doing well in mathematics. Statistically significant differences of the percentages for first- and second-generation immigrant students compared to native students are indicated. The countries are ordered by the differences between native and first-generation immigrant students, beginning with the more positive results for first-generation immigrant students.

I find three participants with a statistically significantly higher percentage of first-generation immigrant students than native students agreeing or strongly agree-

How much do you agree with these statements about learning mathematics?

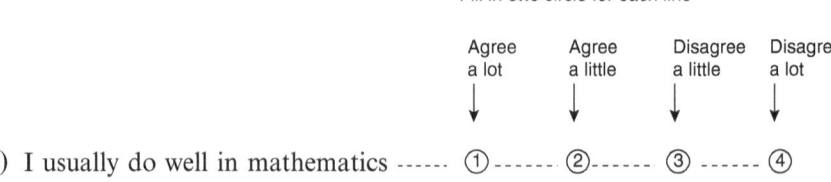

Fig. 4B.7 TIMSS 2007 student questionnaire, variable BS4MAWEL (Source: TIMSS 2007 Assessment. Copyright © 2009 International Association for the Evaluation of Educational Achievement (IEA). Publisher: TIMSS & PIRLS International Study Center, Lynch School of Education, Boston College)

Table 4B.14 Percentages of students who agree or strongly agree to doing well in mathematics by immigration status

Country	First generation immigrant			Native			Second generation immigrant		
	Perc.	SE		Perc.	SE		Perc.	SE	
Korea, Republic of	65	9.5	⇧	44	1.0		12	10.5	⇩
Singapore	81	2.1	⇧	65	1.1		66	1.5	
Armenia	63	3.5		56	1.2		57	3.8	
Canada (British Columbia)	83	2.2	⇧	77	1.2		82	1.4	⇧
Malaysia	49	3.1		45	1.6		43	5.4	
Colombia	84	2.6		80	1.1		77	4.2	
Indonesia	86	1.7		83	1.0		77	6.5	
Czech Republic	67	4.2		63	1.0		63	3.1	
Turkey	81	4.6		78	1.0		81	4.2	
Canada (Ontario)	87	1.4		85	1.4		83	1.5	
Australia	84	2.0		81	1.2		81	1.5	
England	89	2.4		88	0.8		91	1.5	
Qatar	90	0.8		90	0.5		91	0.7	
Jordan	91	1.4		91	0.9		91	1.2	
Mongolia	77	2.1		77	1.3		71	2.7	⇩
United Arab Emirates (Dubai)	83	1.2		84	1.6		82	1.9	
Botswana	75	3.4		75	1.1		70	2.6	
Oman	90	1.3		91	0.6		92	1.3	
Morocco	87	2.8		88	0.7		88	3.5	
Lebanon	86	2.4		86	1.0		83	3.1	
Syria, Arab Republic of	87	1.2		89	0.7		91	1.8	
Sweden	83	2.4		84	0.8		81	1.7	
Bahrain	87	1.9		88	0.7		82	1.9	⇩
Bosnia and Herzegovina	62	1.8		63	1.3		61	4.3	
Ukraine	65	3.9		67	1.6		67	2.7	
Canada (Quebec)	71	2.6		74	1.2		71	2.3	
El Salvador	82	3.2		84	1.0		85	3.4	
Ghana	88	1.2		90	0.9		87	2.5	
Saudi Arabia	89	1.4		91	0.7		91	1.4	
Norway	79	2.6		82	0.7		77	2.1	⇩
Scotland	86	2.4		89	0.7		86	3.4	
Hong Kong, SAR	56	2.0		59	1.7		56	1.6	
Kuwait	86	2.0		89	0.7		89	1.4	
Egypt	90	0.7	⇩	94	0.6		93	1.7	
Russian Federation	57	4.5		61	1.3		62	2.3	
Bulgaria	68	3.8		72	1.3		53	6.0	⇩
Thailand	78	8.0		82	0.8		84	3.9	
Tunisia	83	3.2		87	0.8		89	2.6	
Malta	65	2.7		69	0.7		62	2.3	⇩

(continued)

Table 4B.14 (continued)

Country	First generation immigrant			Native			Second generation immigrant		
	Perc.	SE		Perc.	SE		Perc.	SE	
Israel	81	2.5		86	1.0		86	1.3	
Palestinian National Authority	85	1.5	⇩	90	0.7		88	1.9	
United States (Massachusetts)	81	3.1		87	1.2		82	2.6	
United States	79	1.9	⇩	85	0.7		80	1.2	⇩
Hungary	65	5.0		71	1.1		72	4.4	
Japan	17	7.0		25	1.0		25	6.0	
United States (Minnesota)	77	4.8		85	1.1		81	4.9	
Serbia	54	4.8		63	1.5		65	2.1	
Italy	62	3.7	⇩	70	1.0		66	3.1	
Georgia	68	5.4		78	1.2		71	4.6	
Lithuania	54	4.1	⇩	65	1.0		66	3.0	
Spain (Basque Country)	63	4.6	⇩	73	1.4		66	5.9	
Chinese Taipei	43	3.1	⇩	54	1.2		59	5.6	
Cyprus	67	2.3	⇩	80	0.9		74	2.0	⇩
Slovenia	62	3.5	⇩	75	1.0		69	2.2	⇩
Romania	53	6.7	⇩	70	1.3		62	13.1	
Iran, Islamic Republic of	49	7.3	⇩	77	1.2		71	5.5	

ing to doing well in mathematics: Korea, Singapore, and the Canadian province of British Columbia. As stated above, Korea has very few immigrant students and the statistics are not reliable. For Singapore and British Columbia, the results will be discussed more in-depth in Chaps. 5A and 5B. Especially for British Columbia, the results are remarkable – as we will see in Chap. 5B – because of the high percentage of Asian immigrants which are known for a lower self-rating (see, e.g., Shen & Tam (2008)). On the other hand, I find 11 countries with a statistically significantly lower percentage of students who agree or strongly agree to doing well in mathematics among first-generation immigrant students compared to native students. Calculating the percentage of students who agree or strongly agree to doing well in mathematics, I determine 76 % for native students but only 74 % for first-generation immigrant students.

There are nine countries which display a statistically significantly lower percentage of second-generation immigrant students compared to native students who agree or strongly agree to doing well in mathematics, but none with a statistically significantly higher percentage. As for first-generation immigrant students, the average percentage of students agreeing or strongly agreeing to doing well in mathematics is 74 % – compared to 76 % for native students.

An interesting aspect is to investigate how the differences in self-rating mathematics abilities match the actual differences in mathematics achievement in TIMSS 2007. Table 4B.15 shows the mathematics achievement results for native students and the differences for first- and second-generation immigrant students. Statistically significant differences are indicated. The table also includes the percentage of students agreeing or strongly agreeing to doing well in mathematics for the native students and the differences for first- and second-generation immigrant students. Again, statistically significant differences are indicated.

What we can conclude from Table 4B.15 is, for example, that in Dubai where second-generation immigrants are scoring 65 points higher in mathematics compared with native students and where first-generation immigrants score even 91 score points higher than native students, 84 % of native students answered that they are doing well in mathematics and a slight – not statistically significant – percentage of first- and second-generation immigrant students answered that they are doing

Table 4B.15 Mathematics achievement and students reporting doing well in mathematics for native students and the differences for immigrant students

Country	Native		Difference of immigrants				Natives doing well in math		Differences for immigrants			
	Math Ach	SE	1st gen		2nd gen		Perc.	SE	1st gen		2nd gen	
United Arab Emirates (Dubai)	397	5.7	91	⇧	65	⇧	84	1.6	−0.4		−1.5	
Korea, Republic of	598	2.7	36		−66		44	1.0	21.8	⇧	−31.3	⇩
Canada (British Columbia)	499	2.7	35	⇧	18	⇧	77	1.2	6.3	⇧	5.4	⇧
Singapore	588	3.9	34	⇧	9		65	1.1	15.2	⇧	0.6	
Canada (Ontario)	512	4.5	22	⇧	9		85	1.4	2.4		−1.5	
Armenia	498	2.9	9		5		56	1.2	7.2		0.7	
Australia	496	3.8	1		4		81	1.2	2.3		−0.7	
Bosnia and Herzegovina	457	2.9	−4		5		63	1.3	−1.5		−2.4	
Qatar	305	2.1	−9	⇩	26	⇧	90	0.5	0.2		1.0	
Russian Federation	515	3.9	−15		−12		61	1.3	−3.7		1.0	
Tunisia	423	2.5	−17	⇩	−29	⇩	87	0.8	−4.1		1.4	
Canada (Quebec)	531	3.2	−17		−1		74	1.2	−2.3		−2.4	
Czech Republic	505	2.5	−19	⇩	−14	⇩	63	1.0	3.4		−0.2	
England	515	5.1	−20		13		88	0.8	1.3		3.1	
Botswana	369	2.3	−22	⇩	−28	⇩	75	1.1	−0.4		−5.1	
Hong Kong, SAR	579	5.9	−26	⇩	4		59	1.7	−3.4		−3.0	
Bahrain	403	1.9	−26	⇩	3		88	0.7	−1.3		−6.0	⇩
Israel	469	4.1	−27	⇩	11		86	1.0	−4.9		−0.6	
Iran, Islamic Republic of	405	4.1	−29		−38	⇩	77	1.2	−28.5	⇩	−6.9	
Norway	474	2.2	−29	⇩	−12	⇩	82	0.7	−2.9		−5.1	⇩
Turkey	434	4.8	−29		−30		78	1.0	2.8		3.0	

(continued)

Table 4B.15 (continued)

Country	Native Math Ach	SE	Difference of immigrants 1st gen		2nd gen		Natives doing well in math Perc.	SE	Differences for immigrants 1st gen		2nd gen	
Italy	481	3.2	−30	⇩	−1		70	1.0	−8.6	⇩	−4.2	
Serbia	488	3.5	−31	⇩	8		63	1.5	−8.2		2.5	
Kuwait	362	2.7	−33	⇩	2		89	0.7	−3.6		−0.2	
Japan	571	2.4	−34		−10		25	1.0	−7.8		−0.2	
Lebanon	459	4.0	−36	⇩	−6		86	1.0	−0.7		−3.2	
Saudi Arabia	335	3.0	−37	⇩	16	⇧	91	0.7	−2.5		−0.8	
Scotland	491	3.6	−41	⇩	8		89	0.7	−3.4		−3.6	
Sweden	499	2.2	−41	⇩	−16	⇩	84	0.8	−1.1		−2.9	
Cyprus	472	1.7	−44	⇩	−6		80	0.9	−13.4	⇩	−6.6	⇩
Syria, Arab Republic of	409	3.4	−45	⇩	−23	⇩	89	0.7	−1.1		2.4	
Jordan	430	4.7	−45	⇩	21	⇧	91	0.9	0.1		0.3	
Georgia	418	6.0	−47	⇩	−55	⇩	78	1.2	−10.1		−6.9	
El Salvador	344	2.9	−47	⇩	−3		84	1.0	−2.3		1.1	
United States	517	2.8	−49	⇩	−19	⇩	85	0.7	−6.1	⇩	−5.0	⇩
Malaysia	479	5.0	−50	⇩	−27	⇩	45	1.6	4.6		−1.3	
Palestinian National Authority	380	3.5	−51	⇩	−12		90	0.7	−5.5	⇩	−1.9	
United States (Minnesota)	539	4.1	−53	⇩	−24	⇩	85	1.1	−7.8		−4.0	
Spain (Basque Country)	505	2.8	−57	⇩	−13		73	1.4	−10.7	⇩	−7.1	
Indonesia	409	3.7	−58	⇩	−61	⇩	83	1.0	3.7		−5.7	
Hungary	519	3.3	−58	⇩	4		71	1.1	−6.7		1.0	
Mongolia	448	3.8	−58	⇩	−52	⇩	77	1.3	−0.1		−6.3	⇩
Ghana	324	4.3	−60	⇩	−24	⇩	90	0.9	−2.5		−3.6	
United States (Massachusetts)	558	4.0	−61	⇩	−25	⇩	87	1.2	−6.0		−4.6	
Slovenia	509	2.3	−61	⇩	−21	⇩	75	1.0	−13.6	⇩	−6.3	⇩
Thailand	443	4.9	−62	⇩	−28		82	0.8	−4.0		1.6	
Malta	494	1.5	−62	⇩	−2		69	0.7	−4.4		−6.8	⇩
Lithuania	510	2.3	−64	⇩	−1		65	1.0	−10.7	⇩	1.2	
Bulgaria	473	4.8	−65	⇩	−10		72	1.3	−4.0		−19.5	⇩
Morocco	388	2.9	−65	⇩	−28	⇩	88	0.7	−0.6		0.4	
Colombia	385	3.5	−67	⇩	−41	⇩	80	1.1	3.8		−3.6	
Ukraine	466	3.6	−69	⇩	12	⇧	67	1.6	−2.0		0.1	
Oman	387	3.4	−70	⇩	−20	⇩	91	0.6	−0.6		1.5	
Egypt	427	3.3	−76	⇩	−67	⇩	94	0.6	−3.6	⇩	−0.8	
Romania	466	4.1	−85	⇩	−94	⇩	70	1.3	−16.8	⇩	−7.4	⇩
Chinese Taipei	606	4.3	−111	⇩	−15		54	1.2	−11.0	⇩	5.7	

well in mathematics. Also in Ontario the higher mathematics achievement of first-generation immigrant students is not reflected in higher self-esteem compared to native students, as well as for second-generation immigrant students in Qatar and Saudi Arabia. Based on these results, one might hypothesize that there is a cultural effect in some of the Gulf States that the self-esteem of immigrant students is – compared to their achievement – lower than that of native students in these countries.

In Singapore and British Columbia, on the other hand, the higher achievement of the immigrant populations is reflected in higher self-esteem.

In general I observe in some countries lower self-esteem and also lower mathematics achievement for immigrant students. In some of the countries where immigrant students achieve a more positive outcome, their self-esteem is also higher – but interestingly not in all countries. Especially in the Gulf States, native students seem to have a higher self-esteem than immigrant students, which does not match the achievement results.

Summary

In this chapter I focused on TIMSS 2007 student level data. I aimed to find answers to differences between immigrant and native students with respect to their demographic background.

In this chapter I found that more than half of the countries' first-generation immigrant students are statistically significantly older than their native peers. The same holds true for second-generation immigrant students in more than 20 % of the countries. I could not come to a clear conclusion regarding the relationship between the age of migration and mathematics achievement although I found that in 19 countries the mathematics achievement declined with the age of migration.

I also took into account the sex of the students and its relation to enrollment and achievement. With respect to the enrollment, I found fewer girls than boys enrolled in schools in the majority of countries. With respect to the achievement, I found that first-generation immigrant boys are lagging behind in mathematics more often and to a greater extent than first-generation immigrant girls. For second-generation immigrant boys and girls, I detected the same tendency although less dominant.

As for the language spoken at home, I found many countries with more immigrant students than native students who do not speak the language of instruction at home. But the effect of not speaking the language of instruction at home on the mathematics achievement is larger for native students than for first- and second-generation immigrant students.

I also examined two indicators of the socioeconomic status of the students. I did not find major differences in the parents' education between first- and second-generation immigrant students and native students. I even found that first-generation immigrant students have better educated parents than native students in a couple of countries. The second indicator that I examined was the number of books at home.

In several countries first-generation immigrant students have fewer books at home than native students. I also found the number of books at home being clearly related to the students' mathematics achievement for all groups of students.

Finally in this chapter I addressed students' attitudes towards school in general, the attitudes towards mathematics in particular, and the students' self-esteem in mathematics. For most countries I found no differences in attitudes towards school between immigrant students and native students. For some countries there were differences – sometimes in favor of immigrant students, sometimes in favor of native students.

Regarding the attitude towards mathematics, I found the same result as for the attitudes towards school: for most countries there are no differences between native students and immigrant students, but there are some countries where there are more positive attitudes towards mathematics among native students and among immigrant students. When comparing boys' and girls' attitudes towards mathematics, I couldn't find general patterns among any of the student groups (natives, first-, and second-generation immigrants) that show a higher percentage of boys than girls who like mathematics. Although I detected some differences between countries, there was no general pattern.

For the self-esteem in mathematics, I found a number of countries in which especially first-generation immigrants showed lower self-esteem in mathematics than native students. The difference in self-esteem follows the difference in mathematics achievement in several countries.

In summary, I conclude that immigrant students tend to be older than their native peers, have less affluent – although similarly educated – parents, speak the language of instruction less at home, and have lower self-esteem. First-generation immigrant girls seem to be excluded from education in several countries, but on the other hand, achievement differences for immigrant students in schools are more pronounced for boys.

Despite these general results, I found some countries that have shown more positive results for immigrant students than others. Quite consistently I find Singapore and the Canadian provinces among the countries with positive outcomes for immigrant students.

Chapter 4C
School Factors

Abstract This chapter analyzes differences between schools attended by native students and by immigrant students, taking into account differences between rural and urban schools that can lead to achievement differences, school attendance, school resources, school climate, and school safety. School attendance is associated with mathematics achievement, and that association is stronger for first-generation immigrant students in some countries. Immigrant students and native students are in general attending similar resourced schools, and good resources prove to be positively related to the students' achievement in some countries, as does a positive school rating. However, there are major differences between immigrant and native students in terms of school safety, which is of importance given the clear positive relation between feeling safe in school and mathematics achievement. The results indicate that measures must be found to improve the situation for immigrant students in terms of feeling safer in school.

Keywords School attendance • School resources • School safety

In this chapter, I will analyze if I can find differences between schools attended by native students and by immigrant students. For this I will study differences between rural or urban schools, school attendance, school resources, school climate, and school safety.

Distribution of Immigrants Within Countries

The countries' immigrant student populations are not distributed equally within the countries. For policy makers, it makes a difference if students in more rural areas need special support or if students in urban areas need more support. In rural areas it might be more difficult to access support, such as centers for special language courses, than in urban areas due to a less developed infrastructure. As I will discuss in the next paragraph, achievement differences can also be observed for the different community types.

© Springer International Publishing Switzerland 2016
D. Hastedt, *Mathematics Achievement of Immigrant Students*,
DOI 10.1007/978-3-319-29311-0_6

Table 4C.1 Mathematics achievement in different types of communities

	500.000 and more		3001–500.000		Less than 3000	
Country	math ach	SE	math ach	SE	math ach	SE
Native students	480	2.4	461	0.6	439	1.6
1st gen immigrants	447	4.1	421	1.9	396	3.7
2nd gen immigrants	468	3.1	449	1.6	419	3.8

In TIMSS 2007 the school principals were asked about the type of community in which the school is located. The response options were "3,000 people or fewer", "3,001 to 15,000 people", "15,001 to 50,000 people", "50,001 to 100,000 people", "100,001 to 500,000 people", and "more than 500,000 people". To make the results easier to interpret, the response options, with the exception of the first and last one, were combined. This leaves the response options "3,000 people or fewer", "3,001 to 500,000 people", and "more than 500,000 people". The principal data was matched to the student data and the percentage of immigrant students in the different community types was calculated.

When calculating averages for these three groups of communities in all countries, it becomes clear that the achievement is the lowest in the most rural areas, higher in the mid-sized communities, and the highest in the largest communities. Table 4C.1 shows the mean achievement across all countries participating in TIMSS 2007.

The following two Tables 4C.2 and 4C.3 explain the distribution of first- and second-generation immigrant students within the countries in TIMSS 2007. The data also indicates if the percentage in the extreme categories differs from the percentage in the middle category.

The data allows for the following conclusions. Whereas in some countries there are more immigrant students in more rural areas, there are more countries where there are significantly more immigrant students in the more urban areas. In Australia; Bahrain; the three Canadian provinces British Columbia, Ontario, and Quebec; Japan; Norway; and Dubai, there are statistically significantly more first-generation immigrant students in cities with more than 500,000 inhabitants than in mid-sized areas of 3,001–500,000 inhabitants. In Bahrain, the Canadian provinces Ontario and Quebec, Korea, Tunisia, Dubai, and the United States, there are also statistically significantly more first-generation immigrant students in the mid-sized communities of 3,001–500,000 inhabitants than in the rural communities with less than 3,000 inhabitants.

On the other hand, there are Armenia, Bulgaria, Columbia, Hong Kong, Qatar, and Tunisia where there are statistically significantly more first-generation immigrant students in the mid-sized communities than in the large cities and Bulgaria and Mongolia with statistically significantly more first-generation immigrant students in the rural communities than in the mid-sized communities. The most extreme case is Ontario where I find 29 % of first-generation immigrant students in cities with more than 500,000 inhabitants or more, only 9 % in areas with less than 500,000 but more than 3,000 inhabitants, and only 5 % in areas with less than 3,000 inhabitants.

Table 4C.2 Percentage of first-generation immigrant students in different community types

Country	500.000 and more			3001–500.000			less than 3000	
	Percent	SE	Sign	Percent	SE	Sign	Percent	SE
Canada (Ontario)	29	2.8	⇧	9	1.2	⇧	5	0.9
Canada (British Columbia)	28	3.4	⇧	14	1.6		9	3.3
Bosnia and Herzegovina	38	24.6		23	1.5		28	5.7
United Arab Emirates (Dubai)	58	1.4	⇧	45	4.0	⇧	15	6.9
United States (Minnesota)	18	7.8		7	2.0		4	1.3
Singapore	11	0.6						
Canada (Quebec)	14	1.9	⇧	6	0.8	⇧	2	0.9
United States (Massachusetts)	17	8.0		9	1.1		15	15.1
Australia	15	1.7	⇧	9	1.0		5	2.1
Bahrain	20	2.0	⇧	14	0.8	⇧	11	1.4
Norway	10	1.2	⇧	6	0.4		3	2.6
Czech Republic	7	2.8		2	0.4		2	0.5
United States	13	2.1		9	0.6	⇧	4	1.0
Jordan	16	4.7		12	1.1		12	2.7
Ghana	21	3.0		17	1.7		18	2.7
Spain (Basque Country)	9	8.0		6	0.6			
England	11	2.6		7	0.7		6	1.1
Sweden	10	3.7		8	0.6		5	1.4
Cyprus	13	2.3		10	0.6		9	1.6
Ukraine	9	1.4		7	0.6		8	1.5
Malaysia	10	2.2		8	0.7		6	2.8
Scotland	8	2.8		6	0.7		6	2.6
Indonesia	17	2.3		15	1.4		25	13.3
Lithuania	6	0.9		4	0.4		6	1.2
Morocco	8	1.5		7	0.8		7	9.0
Saudi Arabia	18	2.0		17	1.8		13	2.5
Japan	2	0.4	⇧	1	0.2			
Romania	4	1.4		3	0.5		5	1.2
Italy	6	1.4		5	0.4		4	1.7
El Salvador	6	1.4		5	0.6		7	1.2
Russian Federation	8	1.7		7	0.8		9	2.4
Korea, Republic of	1	0.2		0	0.1			
Serbia	7	1.3		7	0.6		7	2.0
Thailand	1	0.2		1	0.1		0	0.2
Turkey	2	0.4		1	0.3		1	0.9
Iran, Islamic Republic of	1	0.2		1	0.3		0	0.4
Chinese Taipei	6	0.8		6	0.5			
Hungary	2	1.0		3	0.4		4	1.6

(continued)

Table 4C.2 (continued)

Country	500.000 and more		Sign	3001–500.000		Sign	less than 3000	
	Percent	SE		Percent	SE		Percent	SE
Botswana	4	1.8		5	0.5		5	0.8
Israel	12	6.0		13	1.4		19	4.3
Kuwait	15	5.1		17	1.0		13	3.0
Slovenia	3	3.6		5	0.4		5	1.0
Georgia	4	0.7		6	1.0		5	1.2
Egypt	40	4.1		42	2.2		42	6.0
Colombia	3	0.6		6	0.8		6	1.1
Oman	12	4.3		15	1.6		16	1.9
Tunisia	2	1.3		5	0.4	⇧	3	0.8
Syria, Arab Republic of	22	2.6		25	1.3		24	2.7
Lebanon	17	3.5		21	1.8		22	5.8
Bulgaria	6	1.5		10	1.0		15	2.1
Palestinian National Authority	17	3.1		21	1.3		23	5.1
Mongolia	9	2.3		13	1.0		22	3.1
Hong Kong, SAR	22	1.7		27	1.7			
Armenia	7	1.2		13	1.5		11	1.5
Malta				7	0.4		8	1.3
Qatar	18	2.0		26	0.7		29	1.6

Table 4C.3 Percentage of second-generation immigrant students in different community types

Country	500.000 and more		Sign	3001–500.000		Sign	Less than 3000	
	Percent	SE		Percent	SE		Percent	SE
United States (Massachusetts)	40	18.3		17	1.6		10	9.9
Canada (British Columbia)	47	3.8	⇧	25	1.3		19	3.8
Canada (Quebec)	30	4.2	⇧	10	1.2		9	5.5
Singapore	19	0.5	⇧					
Norway	25	3.6	⇧	9	0.6			
England	26	4.1	⇧	12	1.1		11	2.9
Sweden	29	3.0	⇧	16	1.2	⇧	7	2.4
Slovenia	28	11.4		16	1.1	⇧	11	1.7
Bosnia and Herzegovina	16	13.8		5	0.5	⇧	2	0.6
Canada (Ontario)	39	2.9	⇧	28	2.3	⇧	18	2.7
Serbia	27	3.8	⇧	16	1.0		11	2.5
Ukraine	26	2.4	⇧	17	1.0	⇧	12	1.7
Australia	36	1.8	⇧	28	1.4	⇧	10	2.5
United States	25	3.8	⇧	17	1.4	⇧	7	1.6
Jordan	27	2.9		22	1.1	⇧	12	3.4
Saudi Arabia	14	1.7	⇧	10	1.2		8	1.1

(continued)

Table 4C.3 (continued)

Country	500.000 and more			3001–500.000			Less than 3000	
	Percent	SE	Sign	Percent	SE	Sign	Percent	SE
Malaysia	8	1.8		4	0.5		11	5.8
Lithuania	9	1.9		6	0.5		6	1.1
Mongolia	13	3.0		11	0.9		8	1.7
United States (Minnesota)	12	4.1		9	1.1	⇧	3	1.1
Lebanon	8	1.4		6	0.6		4	1.2
Georgia	6	1.5		4	0.8		4	0.9
Scotland	9	2.3		7	0.7		9	1.3
Czech Republic	9	2.0		7	0.5	⇧	4	0.9
Turkey	3	0.7	⇧	1	0.2		2	0.8
Indonesia	2	0.6	⇧	1	0.2		5	0.4
Bulgaria	3	1.0		2	0.2		2	0.6
Japan	2	0.5	⇧	1	0.2			
Colombia	3	0.6		2	0.4		2	0.6
Palestinian National Authority	8	1.5		8	0.6	⇧	4	1.1
Thailand	2	0.7		2	0.4		1	0.6
Romania	1	0.6		0	0.1		1	0.5
Chinese Taipei	3	0.5		2	0.3			
Syria, Arab Republic of	6	0.8		6	0.5		6	1.1
Russian Federation	10	1.0		10	0.8		10	1.7
Iran, Islamic Republic of	2	0.6		2	0.4		2	0.7
Ghana	5	0.7		5	0.7		5	0.6
Hungary	3	1.1		3	0.5		4	0.8
Korea, Republic of	0	0.1		0	0.2			
El Salvador	3	1.1		3	0.4		3	0.8
Bahrain	13	1.5		13	0.7		11	1.2
Tunisia	3	0.8		4	0.3		4	1.1
Armenia	6	1.2		7	0.7	⇧	4	0.9
Kuwait	13	3.5		14	0.9	⇧	11	1.1
Morocco	5	1.0		6	0.7		9	8.1
Egypt	2	0.5		4	0.4		3	0.9
Botswana	9	2.7		10	0.7		11	1.2
Italy	6	1.3		7	0.6		6	1.9
Oman	8	3.2		10	0.8		8	1.4
Spain (Basque Country)	3	2.6		5	0.5		6	3.9
Israel	25	6.1		27	1.1	⇧	5	2.6
Cyprus	7	3.2		10	0.5		9	2.8
Hong Kong, SAR	33	1.7		36	1.2			
Qatar	19	2.4		25	0.6		24	1.3
United Arab Emirates (Dubai)	28	1.1		37	4.6		38	3.7
Malta				12	0.6		12	1.4

This might have an impact on policies addressing immigrant issues in the different countries. While in a larger number of participants first-generation immigrant students are concentrated in the most urban areas – as in Ontario – where, for example, language courses can easily be organized, in a country like Bulgaria, this might be more difficult and more resource intensive. But there are also effects when immigrant students cluster together in classes, as to be discussed later in Chap. 4D.

For second-generation immigrant students, the picture is somewhat more uniform. In 14 countries, there are statistically significantly more students in the most urban areas compared to the mid-sized communities; moreover, there are another 14 countries where there is a statistically significantly lower percentage of second-generation immigrant students in the most rural areas than in the mid-sized communities.

There are only two participants – Qatar and Dubai – where there is a statistically significantly lower percentage of second-generation immigrant students in the cities with more than 500,000 inhabitants than in the mid-sized communities, and only in Indonesia there is a statistically significantly higher percentage of second-generation immigrant students in the communities with less than 3,000 inhabitants than in the mid-sized communities. The highest difference in the percentages can be seen in British Columbia where I find 47 % of second-generation immigrant students in cities with 500,000 inhabitants or more, but only 25 % in areas with less than 500,000 inhabitants but more than 3,000 inhabitants and only 19 % in areas with less than 3,000 inhabitants.

But does the community size have an impact on the differences between native students and immigrant students? Next I calculated the difference of mathematics achievement between first- and second-generation immigrant students and their native peers in the three types of communities for each participating country.

Table 4C.4 shows the mathematics achievement difference between first-generation immigrant students and native students in cities with 500,000 inhabitants or more, in communities with 3,000–500,000 inhabitants, and in rural areas with less than 3,000 inhabitants. Again, the picture is quite diverse, but I find some very interesting results. In Hungary, for example, first-generation immigrant students in the big cities with 500,000 or more inhabitants outperform their native peers in the cities by 45 score points, whereas in the mid-sized communities with less than 500,000 inhabitants but more than 3,000 inhabitants, native students outperform their first-generation immigrant peers by 51 score points. In the rural communities with less than 3,000 inhabitants, this difference in favor of native students even increases to 83 score points. Also in Ontario, Jordan, and Slovenia, I see similar effects between the cities with 500,000 inhabitants and more and the mid-sized communities.

Table 4C.5 shows the mathematics achievement difference between second-generation immigrant students and native students in the cities with 500,000

Table 4C.4 Mathematics achievement of differences between first-generation immigrant students and native students in different types of communities

Country	500.000 and more Difference immi-native	SE	Sign	3001–500.000 Difference immi-native	SE	Sign	Less than 3000 Difference immi-native	SE
Korea, Republic of	59	23.0		−40	45.4			
Hungary	45	30.2	⇑	−51	19.3		−83	45.0
Malta				−68	6.6		−34	17.3
Jordan	−4	18.5	⇑	−62	11.1		−59	25.9
Chinese Taipei	−66	26.3		−121	10.6			
Bulgaria	−20	26.5		−64	11.9		−56	24.5
Israel	13	31.5		−32	10.0		−35	27.3
Slovenia	−20	15.1	⇑	−60	9.0		−73	14.0
Thailand	−32	43.7		−71	36.6		−105	57.8
United States (Minnesota)	−28	11.0		−64	15.2		−22	24.2
Singapore	34	7.2						
England	8	28.2		−23	13.6		−28	73.1
Georgia	−21	22.8		−51	19.7		−46	29.7
Canada (Ontario)	27	9.9	⇑	−2	10.7		2	19.6
Mongolia	−31	26.3		−60	7.9		−54	12.7
Serbia	−6	24.3		−35	10.7		−31	29.8
Italy	−8	18.1		−30	7.0		−66	37.4
Palestinian National Authority	−28	21.6		−49	8.3		−70	31.4
Ukraine	−48	18.5		−69	12.2		−86	14.5
Hong Kong, SAR	−7	13.3		−27	12.0			
Tunisia	2	16.6		−17	6.9		−4	18.5
Lithuania	−57	23.7		−72	9.3		−50	21.4
Norway	−18	14.9		−29	5.4		−34	43.8
United Arab Emirates (Dubai)	94	12.2		83	11.1	⇑	−8	37.2
Sweden	−33	20.7		−43	7.4		−24	32.9
Lebanon	−24	32.0		−34	7.8		−53	15.8
Australia	0	15.5		−9	12.6		−23	16.5
United States (Massachusetts)	−56	91.1		−65	10.3		−10	9.1
Iran, Islamic Republic of	−29	36.7		−36	26.1		7	23.4
Indonesia	−51	23.1		−58	7.4		−39	22.3
Malaysia	−46	24.8		−52	10.4		−60	34.9
El Salvador	−36	35.1		−40	14.6		−53	16.1
United States	−51	17.2		−53	7.6		−62	36.7
Scotland	−39	34.6		−36	14.4		−67	13.0

(continued)

Table 4C.4 (continued)

Country	500.000 and more			3001–500.000			Less than 3000	
	Difference immi- native	SE	Sign	Difference immi- native	SE	Sign	Difference immi- native	SE
Colombia	−69	15.4		−66	11.8		−39	27.9
Oman	−71	30.5		−67	9.3		−73	14.9
Bahrain	−30	10.6		−25	5.9		1	11.9
Morocco	−69	14.4		−64	10.1		−7	20.2
Canada (Quebec)	−27	16.9		−21	12.4		−43	41.6
Egypt	−81	14.5		−74	7.8		−65	16.3
Syria, Arab Republic of	−48	9.6		−42	9.7		−50	15.7
Romania	−92	26.7		−85	13.7		−73	26.5
Cyprus	−54	30.1		−42	6.3		−58	16.8
Ghana	−70	17.0		−55	11.1		−60	12.7
Japan	−41	21.0		−25	34.0			
Saudi Arabia	−45	11.9		−27	11.0	⇧	−64	14.3
Czech Republic	−43	26.5		−23	10.1		9	17.6
Russian Federation	−35	15.1		−14	11.3		4	18.1
Bosnia and Herzegovina	−29	89.8		−4	6.0		−5	11.7
Canada (British Columbia)	15	20.4		41	9.8		−6	41.9
Spain (Basque Country)	−93	113.4		−55	9.8			
Armenia	−25	14.7		18	19.5		6	19.6
Turkey	−68	28.6		−20	25.5		10	65.7
Kuwait	−81	61.8		−30	6.9		−25	23.9
Qatar	−65	15.2		−10	4.1		7	7.6
Botswana	−76	35.3		−9	13.5		−33	12.7

inhabitants or more, in communities with 3,000–500,000 inhabitants, and in rural areas with less than 3,000 inhabitants. The table suggests that in Cyprus, second-generation immigrant students are very well-off in big cities compared to native students. However this statistic is based on three second-generation immigrant students, so this result is not reliable and I will ignore it. Also in Tunisia, the statistic is based on only five second-generation immigrant students in cities with 500,000 inhabitants and more. Also for several other countries, the big standard errors indicate very small sample sizes and consequently not very reliable statistics. Actually, I do not find reliable statistical differences for second-generation immigrants and native students between the different community types.

Table 4C.5 Mathematics achievement differences between second generation immigrant students and native students in different types of communities

Country	500.000 and more			3001–500.000			less than 3000	
	Difference immi-native	SE	sign	Difference immi-native	SE	sign	Difference immi-native	SE
Korea, Republic of	59	23.0	0.0	−40	45.4	0.0		
Hungary	45	30.2	1.0	−51	19.3	0.0	−83	45.0
Malta				−68	6.6	0.0	−34	17.3
Jordan	−4	18.5	1.0	−62	11.1	0.0	−59	25.9
Chinese Taipei	−66	26.3	0.0	−121	10.6	−1.0		
Bulgaria	−20	26.5	0.0	−64	11.9	0.0	−56	24.5
Israel	13	31.5	0.0	−32	10.0	0.0	−35	27.3
Slovenia	−20	15.1	1.0	−60	9.0	0.0	−73	14.0
Thailand	−32	43.7	0.0	−71	36.6	0.0	−105	57.8
United States (Minnesota)	−28	11.0	0.0	−64	15.2	0.0	−22	24.2
Singapore	34	7.2						
England	8	28.2	0.0	−23	13.6	0.0	−28	73.1
Georgia	−21	22.8	0.0	−51	19.7	0.0	−46	29.7
Canada (Ontario)	27	9.9	1.0	−2	10.7	0.0	2	19.6
Mongolia	−31	26.3	0.0	−60	7.9	0.0	−54	12.7
Serbia	−6	24.3	0.0	−35	10.7	0.0	−31	29.8
Italy	−8	18.1	0.0	−30	7.0	0.0	−66	37.4
Palestinian National Authority	−28	21.6	0.0	−49	8.3	0.0	−70	31.4
Ukraine	−48	18.5	0.0	−69	12.2	0.0	−86	14.5
Hong Kong, SAR	−7	13.3	0.0	−27	12.0	−1.0		
Tunisia	2	16.6	0.0	−17	6.9	0.0	−4	18.5
Lithuania	−57	23.7	0.0	−72	9.3	0.0	−50	21.4
Norway	−18	14.9	0.0	−29	5.4	0.0	−34	43.8
United Arab Emirates (Dubai)	94	12.2	0.0	83	11.1	1.0	−8	37.2
Sweden	−33	20.7	0.0	−43	7.4	0.0	−24	32.9
Lebanon	−24	32.0	0.0	−34	7.8	0.0	−53	15.8
Australia	0	15.5	0.0	−9	12.6	0.0	−23	16.5
United States (Massachusetts)	−56	91.1	0.0	−65	10.3	−1.0	−10	9.1
Iran, Islamic Republic of	−29	36.7	0.0	−36	26.1	0.0	7	23.4
Indonesia	−51	23.1	0.0	−58	7.4	0.0	−39	22.3
Malaysia	−46	24.8	0.0	−52	10.4	0.0	−60	34.9
El Salvador	−36	35.1	0.0	−40	14.6	0.0	−53	16.1
United States	−51	17.2	0.0	−53	7.6	0.0	−62	36.7
Scotland	−39	34.6	0.0	−36	14.4	0.0	−67	13.0
Colombia	−69	15.4	0.0	−66	11.8	0.0	−39	27.9

(continued)

Table 4C.5 (continued)

Country	500.000 and more			3001–500.000			less than 3000	
	Difference immi-native	SE	sign	Difference immi-native	SE	sign	Difference immi-native	SE
Oman	−71	30.5	0.0	−67	9.3	0.0	−73	14.9
Bahrain	−30	10.6	0.0	−25	5.9	−1.0	1	11.9
Morocco	−69	14.4	0.0	−64	10.1	−1.0	−7	20.2
Canada (Quebec)	−27	16.9	0.0	−21	12.4	0.0	−43	41.6
Egypt	−81	14.5	0.0	−74	7.8	0.0	−65	16.3
Syria, Arab Republic of	−48	9.6	0.0	−42	9.7	0.0	−50	15.7
Romania	−92	26.7	0.0	−85	13.7	0.0	−73	26.5
Cyprus	−54	30.1	0.0	−42	6.3	0.0	−58	16.8
Ghana	−70	17.0	0.0	−55	11.1	0.0	−60	12.7
Japan	−41	21.0	0.0	−25	34.0	0.0		
Saudi Arabia	−45	11.9	0.0	−27	11.0	1.0	−64	14.3
Czech Republic	−43	26.5	0.0	−23	10.1	0.0	9	17.6
Russian Federation	−35	15.1	0.0	−14	11.3	0.0	4	18.1
Bosnia and Herzegovina	−29	89.8	0.0	−4	6.0	0.0	−5	11.7
Canada (British Columbia)	15	20.4	0.0	41	9.8	0.0	−6	41.9
Spain (Basque Country)	−93	113.4	0.0	−55	9.8	−1.0		
Armenia	−25	14.7	0.0	18	19.5	0.0	6	19.6
Turkey	−68	28.6	0.0	−20	25.5	0.0	10	65.7
Kuwait	−81	61.8	0.0	−30	6.9	0.0	−25	23.9
Qatar	−65	15.2	−1.0	−10	4.1	0.0	7	7.6
Botswana	−76	35.3	0.0	−9	13.5	0.0	−33	12.7

I conclude from this analysis that immigrant students are concentrated in some countries in more urban areas – especially the second-generation immigrant students – but in some countries also in more rural areas, which poses different challenges to the education systems of the different countries. There is no unique picture but some interesting differences across countries. In terms of achievement, I see for the first-generation immigrant students some differences in some countries. Some countries seem to be able to offer good opportunities for immigrant students mostly in large cities but facing challenges in more rural areas.

School Attendance

Now, I want to take a look at school attendance as the first school factor and consequently as the first part of research question six. As found in the literature review in Chap. 2, school attendance can be a problematic issue in schools and achievement results usually relate positively to school attendance.

Table 4C.6 shows the data for all countries. It is indicated where the percentages for the immigrant groups statistically significantly differ from those for native students. In Cyprus, for example, the percentage of first-generation immigrant students attending schools with low school attendance is 22.8 %, whereas it is only 15.1 % among native students.

Table 4C.6 Percentage of students in schools with low school attendance

Country	First generation immigrant			Native		Second generation immigrant	
	Percent	SE		Percent	SE	Percent	SE
Japan	60.7	10.2	⇧	38.8	3.9	43.0	9.0
Lithuania	53.5	6.4		49.4	4.5	54.5	5.6
Colombia	49.5	6.8		46.1	4.2	51.2	7.0
Bulgaria	48.7	5.3		36.7	4.2	37.7	7.3
Indonesia	44.9	6.1		33.1	4.4	39.0	9.8
Mongolia	43.1	4.9		47.8	4.8	51.8	5.2
Sweden	42.4	5.3		37.9	4.1	34.5	5.2
Serbia	39.7	6.2		26.7	3.5	33.3	4.6
Kuwait	38.8	5.4		35.1	4.4	37.7	5.4
Morocco	38.3	10.8		43.8	6.3	37.5	7.6
Romania	36.6	7.6		29.7	4.1	24.9	9.9
United States	27.5	5.2	⇧	16.2	2.5	26.0	5.9
Georgia	25.5	6.4		21.4	4.3	15.5	4.5
Canada (Quebec)	25.2	5.5		24.8	4.0	23.1	6.0
Ghana	24.5	5.0		24.4	4.1	28.0	5.8
Hungary	23.3	6.7		19.1	3.7	20.9	6.1
Cyprus	22.8	2.4	⇧	15.1	0.4	15.0	1.4
Israel	22.6	5.1		23.4	4.3	24.9	4.7
Russian Federation	21.9	6.3		20.3	2.8	16.8	3.2
Slovenia	21.8	5.0		17.2	3.0	24.6	5.3
Botswana	21.7	4.0		27.9	3.7	20.0	3.4
Malaysia	21.6	5.9		14.8	2.8	16.9	4.9
Jordan	19.6	4.8		17.2	3.3	16.3	3.7
Saudi Arabia	19.3	3.9		20.6	3.3	26.9	4.9
Malta	18.5	2.3	⇧	8.7	0.4	9.8	1.3
Australia	18.3	4.4		15.8	3.1	16.2	3.2
Qatar	17.6	0.8	⇩	27.0	0.5	21.4	0.9

(continued)

Table 4C.6 (continued)

Country	First generation immigrant			Native		Second generation immigrant	
	Percent	SE		Percent	SE	Percent	SE
Spain (Basque Country)	17.6	4.8		7.9	2.6	11.0	3.7
Tunisia	17.5	4.6		22.8	3.8	29.0	6.0
Bahrain	17.3	1.8		20.2	0.5	16.0	1.7
Norway	17.2	4.1		19.2	3.7	18.0	4.2
Syria, Arab Republic of	17.0	3.1		19.6	3.6	22.5	4.5
Palestinian National Authority	16.9	3.9		13.4	2.4	10.7	2.2
El Salvador	16.8	4.2		22.7	3.9	13.1	5.1
Hong Kong, SAR	15.5	4.8		6.7	2.2	9.5	3.0
United States (Massachusetts)	14.5	8.8		7.6	4.3	11.1	6.5
England	14.5	4.7		11.3	2.7	13.5	4.8
Czech Republic	14.4	6.3		10.8	2.8	11.7	3.9
Scotland	13.2	5.6		6.7	1.9	3.8	2.0
Turkey	12.9	4.9		22.3	3.5	25.4	8.2
Ukraine	12.6	4.9		12.5	3.1	12.3	3.9
Armenia	12.6	3.2		13.0	2.5	12.0	4.5
United Arab Emirates (Dubai)	12.0	1.0	⇧	4.7	0.5	5.3	0.7
Canada (British Columbia)	11.9	3.6		22.0	4.1	16.0	3.8
Oman	11.2	4.4		8.0	2.5	11.0	4.2
Egypt	11.1	2.8		13.6	2.9	16.9	5.1
Thailand	10.6	7.2		17.5	3.5	20.4	7.3
Italy	10.1	2.7		16.1	2.8	10.8	2.5
Canada (Ontario)	9.4	4.4		10.5	3.1	7.3	3.0
Bosnia and Herzegovina	8.9	2.0		11.9	3.0	14.4	4.3
Lebanon	6.9	3.3		4.6	1.4	9.1	4.2
Singapore	5.8	1.7		4.0	0.2	3.4	0.7
Korea, Republic of	4.9	5.1		9.2	1.8	9.5	10.0
United States (Minnesota)	4.0	3.5		1.5	0.9	2.1	2.1
Chinese Taipei	3.0	1.5		5.4	1.9	10.8	4.5
Iran, Islamic Republic of	0.0			3.4	1.3	3.7	2.5

Also in Japan, Malta, Dubai, and the United States, the percentage is statistically significantly higher among first-generation immigrant students than among native students. For Japan, it must be noted that the statistics are based on less than 50 first-generation immigrant students and still less than 100 second-generation immigrant students. Consequently, there is a high probability that the result found is an artifact.

For Bahrain and Qatar, the situation is statistically significantly more positive among second-generation immigrant students than among native students, whereas in Qatar the percentage of first-generation immigrant students is even smaller. For several countries, the differences of the percentages, although quite high, are not statistically significantly different due to the large sampling error for this variable. In Serbia, for example, there are 13 % more first-generation immigrant students than native students in the group, but this difference is not statistically significant. It should also be noted here that in most countries there is a lower percentage of immigrant students than native students enrolled in schools with a high school attendance. Only in Singapore there is a statistically significantly higher percentage of first-generation immigrant students (35.7 %) than native students (29.0 %) in schools with high attendance.

But what can also be seen in Table 4C.6 is that overall there is not much difference between immigrant students and native students in terms of how many percent go to schools with high, medium, or low school attendance. Across all countries, the average percentage of native students in schools with low school attendance is 19.4 %, of first-generation immigrant students 21.2 %, and of second-generation immigrant students 20.1 %. This means that in most countries low school attendance is not a problematic issue specifically for immigrant students, even though there is a slight tendency that a higher percentage of first-generation immigrant students goes to schools with low school attendance.

Next I am going to analyze the mathematics achievement differences for students attending schools with high, medium, or low student participation.

Students' school attendance has consequences for their achievement. As can be seen in Exhibit 8.3 in Mullis et al. (2008), students attending schools with a better school attendance performed better in most of the countries. Table 4C.7 shows the achievement of first-generation immigrant students attending any of the three groups of schools.

The results confirm that first-generation immigrant students attending schools with a better school attendance performed better than first-generation immigrant students attending schools with medium or low school attendance in most of the countries. The average mathematics achievement for first-generation immigrant students is 449 points in schools with high school attendance, 429 points in schools with middle school attendance, and 412 points in schools with low school attendance.

Interesting counterexamples are Armenia or Slovenia where students in the highest group performed statistically significant lower than students in the medium group. In Slovenia this might be caused by the small sample with less than 50 students in the two extreme groups of student participation among first-generation immigrant students, which might lead to unstable estimates. In Armenia, however, a similar problem could not be found and we are left with the puzzling result. Interestingly, we find in Dubai that the students in the lowest category performed above the medium group.

Very interesting are the last two columns in Table 4C.7 that compare the differences between the students in the highest group of school attendance and the lowest for first-generation immigrant students and native students. In most countries, the difference is higher for immigrant students than for native students – in some coun-

Table 4C.7 Mathematics achievement of first-generation immigrant students in schools with different levels of school attendance

Country	High Math Ach	SE		Medium Math Ach	SE	Low Math Ach	SE		Difference between high and low groups Math Ach	SE		Difference between high and low for native students Math Ach	SE	
Morocco	429	44.1	⇧	319	9.3	313	10.1		116	45.3		44	24.9	
Malta	511	9.6	⇧	403	8.0	422	14.6		89	17.5		84	4.7	
Lithuania	555	99.7		453	11.5	430	10.8		125	100.3		−20	10.1	
Scotland	523	22.8	⇧	439	12.5	444	35.5		79	42.2		48	23.6	
Mongolia	453	28.0	⇧	385	8.3	390	8.3		63	29.2		10	27.6	
Sweden	528	23.3	⇧	462	8.3	444	10.4		83	25.5		31	17.7	
Canada (Quebec)	564	22.7	⇧	508	10.2	500	13.4		64	26.3		54	12.0	
Australia	554	16.9	⇧	500	8.6	425	11.8	⇩	129	20.6		92	12.7	
Singapore	658	10.8	⇧	604	7.2	560	24.8		97	27.0		94	33.5	
England	538	26.2		490	12.8	460	17.0		77	31.2		74	17.6	
Hong Kong, SAR	596	11.0	⇧	549	7.1	472	21.6	⇩	124	24.2		99	31.0	
Japan	612	21.2		573	29.9	518	26.2		94	33.7		14	9.1	⇧
Botswana	392	32.0		354	10.6	304	17.7	⇩	89	36.6		34	9.1	
United Arab Emirates (Dubai)	508	5.2	⇧	471	5.5	566	6.8	⇧	−58	8.5		−17	23.3	
Russian Federation	532	23.9		496	8.7	492	14.0		40	27.6		37	10.9	
Bahrain	412	6.7	⇧	376	5.6	355	8.3	⇩	57	10.6		21	5.9	⇧
Qatar	321	7.9	⇧	286	3.3	297	6.5		24	10.2		−3	5.8	⇧
Ukraine	427	20.5		393	11.0	385	11.8		42	23.7		31	11.6	
Turkey	446	45.1		412	22.8	332	30.9	⇩	114	54.7		37	16.0	
El Salvador	326	16.0		298	13.1	286	17.3		40	23.5		26	11.9	
Iran, Islamic Republic of	391	69.0		371	22.4	NA			NA			9	10.9	
Bulgaria	438	17.0		419	10.5	392	16.2		45	23.5		41	13.8	
Oman	333	9.2		316	11.4	277	20.4		55	22.4		15	18.0	
Cyprus	443	19.2		429	6.4	416	13.3		27	23.3		−11	7.2	
Malaysia	443	26.2		429	11.0	420	12.0		23	28.8		47	17.8	
Indonesia	377	19.0		363	9.1	334	9.3	⇩	43	21.2		48	19.1	
Egypt	361	9.0		348	7.9	331	8.9		30	12.6		32	12.2	
Kuwait	336	15.5		324	10.8	333	9.9		4	18.4		12	9.7	
United States	485	15.9		473	7.8	443	9.5	⇩	42	18.6		36	8.6	
United States (Minnesota)	501	10.5		490	13.5	369	21.1	⇩	131	23.6		109	13.3	
Canada (Ontario)	546	22.8		536	6.5	500	13.7	⇩	46	26.6		18	17.1	
Romania	391	40.7		382	15.9	375	14.9		16	43.3		39	13.7	
Chinese Taipei	499	14.2		491	13.4	469	33.4		30	36.3		30	12.3	

(continued)

Table 4C.7 (continued)

Country	High Math Ach	SE	Medium Math Ach	SE	Low Math Ach	SE	Difference between high and low groups Math Ach	SE	Difference between high and low for native students Math Ach	SE
Hungary	470	78.2	462	24.1	452	22.7	18	81.4	34	11.4
Tunisia	416	14.9	409	8.6	400	13.6	16	20.1	6	8.0
United States (Massachusetts)	509	41.2	506	12.0	436	15.1 ⇩	73	43.9	48	21.3
Thailand	386	60.1	384	34.8	354	50.2	32	78.4	16	19.8
Bosnia and Herzegovina	453	7.5	451	6.1	457	7.7	−4	10.8	9	13.1
Lebanon	424	8.9	423	10.9	403	19.2	21	21.2	18	16.7
Canada (British Columbia)	544	13.6	544	10.5	502	13.4 ⇩	43	19.1	44	14.5
Norway	443	16.6	445	5.6	454	11.6	−10	20.2	17	8.5
Ghana	270	71.5	272	7.3	238	10.4 ⇩	33	72.3	96	26.3
Saudi Arabia	296	16.3	298	7.7	317	15.8	−21	22.7	−14	10.4
Spain (Basque Country)	447	20.7	449	11.5	456	21.4	−9	29.7	19	12.0
Serbia	446	19.5	449	14.4	463	13.2	−17	23.6	15	10.2
Italy	448	13.4	452	7.0	459	21.2	−10	25.1	12	9.9
Colombia	315	16.8	322	21.3	314	10.3	1	19.7	30	11.6
Israel	436	27.1	443	8.9	449	18.4	−13	32.8	11	13.4
Jordan	382	13.8	390	13.2	365	20.3	17	24.5	27	14.4
Palestinian National Authority	330	15.3	340	8.6	288	18.8 ⇩	41	24.2	43	11.0
Syria, Arab Republic of	358	12.6	368	7.9	350	13.3	8	18.3	−17	10.7
Czech Republic	482	18.5	493	8.6	479	11.3	3	21.7	35	7.9
Korea, Republic of	625	44.9	637	39.0	696	19.9	−71	49.1	1	9.2
Slovenia	426	15.4 ⇩	463	9.4	433	15.8	−8	22.1	−2	6.8
Georgia	335	32.3	378	13.6	366	31.6	−31	45.2	−28	29.0
Armenia	468	11.1 ⇩	526	18.7	481	16.9	−12	20.2	0	7.8

tries, it is even tremendously high. On average I find that the difference for first-generation immigrant students attending schools with high school attendance compared to those attending schools with low school attendance amounts to 38 points compared to only a 29-score-point difference for native students. Following the argument of Büchel et al. (2001) that an important role of the school is also the social aspect, one might conclude that a higher school attendance could be especially important for immigrant students to integrate into the host society, which then would also lead to better achievement.

The case of Lithuania shows a difficulty with these statistics. In Lithuania native students in schools with high school attendance score 20 points below native students in schools with low school attendance, whereas for first-generation immigrant students, the ones in schools with high school attendance score 125 above first-generation immigrant students who attend schools with low school attendance. The large standard errors for the achievement of students in schools with high school attendance – especially for first-generation immigrants – are mainly caused by the low percentage of students in this group, 6.0 % for native students and 3.9 % for first-generation immigrant students plus the huge variance between the students in this group. Finally it can be concluded that there are at least some first-generation immigrant students in Lithuania in schools with low school attendance that considerably underachieved in mathematics. Although further research of this case is necessary, it might be a starting point for finding reasons and measures for improving the achievement of first-generation immigrant students in Lithuania.

We find similar results in Sweden, Botswana, or Morocco with huge differences of the achievement gaps but also huge standard errors for these differences due to very small percentages in the group of students attending schools with high attendance rates. Again, further research is necessary to explore this finding, but the data gives a hint for a starting point in analyzing the difficulties of immigrant students in these countries.

The only countries where the differences in mathematics achievement are statistically significant are Bahrain, Qatar, and Japan. For these countries, we can say that attending a school with high or low school attendance has a statistically significantly higher effect for first-generation immigrant students than for native students. Approaching the problem of low school attendance in schools with a high proportion of immigrant students might help improve the achievement of these students.

I conclude that in terms of school attendance, there is a slightly higher percentage of first-generation immigrant students in schools with low school attendance. I also found that lower school attendance is associated with lower mathematics achievement for first-generation immigrant students in TIMSS 2007. The differences between students in schools with a high or low school attendance are on average higher for first-generation immigrant students than for native students.

School Resources

That adequate school resources play an important role and a shortage can affect the teaching negatively was already discussed in Chap. 2. Hansson and Gustafsson (2010, p. 12) stated: "Neighborhood and school SES, not family SES, may exert a more powerful effect on academic outcomes in minority communities." In TIMSS the school resources for mathematics teaching were captured in the school principal questionnaire. Principals were asked to which degree general instruction was affected by a shortage or inadequacy of instructional materials, budget for supplies, school buildings and grounds, heating/cooling and lightning systems, and instructional space.

Regarding the mathematics teaching, there were similar questions about a shortage of computers, software, calculators, library materials, or audiovisual resources that might affect the mathematics teaching negatively. The response options were none (coded to 1), a little (coded to 2), some (coded to 3), and a lot (coded to 4). Schools were marked as highly resourced if both areas – effects for general instruction and mathematics instruction – had an average below 2 (a little). Details can be found in Mullis et al. (2008, p. 342). As for the achievement, the result was not surprisingly that "Students at the high level of the index had the highest average mathematic achievement (464 points), followed by the students at the medium level (449 points) and then by the students at the low level (420 points)" (Mullis et al., 2008, p. 343).

Table 4C.8 shows the percentage of native students and first- and second-generation immigrant students that attend high-resourced schools. It is also indicated if the percentages for first- or second-generation immigrant students are statistically significantly different from the percentages for native students for this group.

Although on average the percentage is the same for the three groups of students, there are some countries where there is a statistically significant difference – mostly to the disadvantage of immigrant students. The Czech Republic, Qatar, Cyprus, Iran, and Malta are the countries that show a statistically significantly lower percentage of first-generation immigrant students in well-resourced schools than native students. Only in Dubai there is a statistically significantly higher percentage of first-generation immigrant students in well-resourced schools than native students.

Malta and Qatar also show a statistically significantly lower percentage of second-generation immigrant students in well-resourced schools than native students. But on the other hand, Dubai as well as Bahrain and Cyprus show a statistically significantly higher percentage of second-generation immigrant students in well-resourced schools than native students.

When examining the achievement differences, one finds, not surprisingly, that on average students in well-resourced schools achieved better than students attending medium- or low-resourced schools. Tables 4C.9 and 4C.10 show the results for first- and second-generation immigrant students. Statistically significant differences for the well-resourced and low-resourced schools from medium-resourced schools are also indicated. Since some countries had no schools where the principal indicated a low-resourced school for these countries, the achievement is indicated as "NA." The general picture is the same, showing that also immigrant students who attend better resourced schools achieved higher – and in a fair amount of countries statistically significant higher – than their peers in less resourced schools.

For first-generation immigrant students, the highest effect of high-resourced schools can be seen in Thailand where the students in high-resourced schools achieved 175 score points above the students in medium-resourced schools. However, one must take into account that this statistic is based on only 28 first-

Table 4C.8 Percentage of students in high-resourced schools

Country	First generation immigrant			Native		Second generation immigrant		
	Percent	SE		Percent	SE	Percent	SE	
United Arab Emirates (Dubai)	77.6	1.5	⇧	58.8	2.7	75.2	2.1	⇧
Korea, Republic of	38.9	14.3		29.8	3.9	32.0	17.3	
Australia	61.0	5.4		52.4	3.8	58.0	4.7	
Thailand	20.5	10.1		13.0	2.5	13.5	5.3	
England	40.6	6.3		33.2	3.9	32.8	5.4	
Turkey	13.6	5.5		7.5	2.3	15.5	9.0	
Jordan	26.1	8.2		20.4	3.6	19.3	3.5	
Spain (Basque Country)	75.0	5.8		69.5	4.6	65.4	6.8	
Canada (British Columbia)	59.8	6.4		54.7	5.3	57.5	5.4	
Sweden	52.6	6.0		47.8	4.4	52.6	5.6	
Botswana	8.9	3.8		4.1	1.7	5.8	2.3	
Norway	25.4	4.8		20.9	3.7	30.8	6.0	
Ghana	13.8	4.1		9.4	2.6	9.5	3.5	
United States (Minnesota)	52.0	12.9		47.6	9.2	43.9	10.7	
Scotland	51.5	6.0		47.2	4.7	48.7	5.5	
Saudi Arabia	10.2	3.2		7.3	2.1	5.4	1.8	
Italy	27.7	5.1		24.9	3.4	17.7	3.7	
Colombia	18.9	5.9		16.2	3.5	21.5	7.4	
Tunisia	8.2	2.0		5.7	1.7	3.8	1.7	
El Salvador	15.2	3.8		13.1	2.6	11.7	3.4	
Oman	16.9	3.8		15.2	3.1	21.6	4.6	
Ukraine	14.7	5.1		13.2	3.1	11.2	3.0	
Lebanon	37.5	5.9		36.7	4.2	37.3	5.9	
Romania	19.4	6.9		18.7	3.4	19.1	8.3	
Malaysia	43.2	5.1		42.5	4.4	40.9	6.3	
Morocco	2.1	1.5		1.5	0.9	0.3	0.3	
Japan	51.1	10.2		50.8	4.1	57.1	9.5	
Singapore	91.4	1.6		91.2	0.4	91.4	0.8	
Palestinian National Authority	18.9	4.0		19.1	3.2	23.7	4.7	
Chinese Taipei	35.3	4.7		35.8	3.8	34.6	5.7	
Bahrain	23.3	1.9		23.8	0.5	29.2	2.2	⇧
Indonesia	5.6	2.7		6.3	2.1	3.6	2.6	
Bosnia and Herzegovina	5.5	2.0		6.3	1.9	10.1	3.4	
Mongolia	4.8	1.9		6.1	2.3	4.4	2.7	
Lithuania	21.1	5.6		22.5	3.9	22.6	5.6	
Serbia	13.2	4.1		14.9	3.1	16.5	4.1	
Syria, Arab Republic of	11.2	2.7		13.0	2.8	12.9	4.6	
Israel	34.5	7.1		36.4	4.4	38.6	5.3	

(continued)

Table 4C.8 (continued)

Country	First generation immigrant			Native			Second generation immigrant		
	Percent	SE		Percent	SE		Percent	SE	
Georgia	4.6	2.2		6.5	2.3		3.9	1.5	
Egypt	25.4	4.3		27.3	3.9		23.9	4.9	
Canada (Ontario)	35.0	8.7		37.5	5.0		32.0	5.0	
Hungary	47.4	7.8		50.0	4.8		38.0	7.0	
Kuwait	11.1	3.0		13.9	3.1		14.5	3.7	
Cyprus	8.6	1.0	⇩	11.7	0.4		15.3	1.5	⇧
Canada (Quebec)	50.5	6.5		53.8	5.4		54.2	6.5	
Slovenia	59.9	5.6		63.3	4.4		60.5	5.9	
United States	47.9	4.6		52.1	3.8		50.9	5.5	
Armenia	14.5	3.2		19.3	3.4		15.8	4.3	
Bulgaria	24.7	5.2		29.9	3.6		31.1	7.7	
Malta	49.5	2.4	⇩	55.1	0.4		50.1	2.0	⇩
Hong Kong, SAR	68.6	4.7		74.5	3.7		67.2	4.5	
Russian Federation	19.9	3.7		28.3	2.9		33.3	5.0	
Iran, Islamic Republic of	1.9	1.4	⇩	10.8	2.2		6.5	2.8	
Qatar	20.9	0.8	⇩	33.6	0.5		21.5	0.9	⇩
United States (Massachusetts)	37.2	8.8		52.8	7.5		37.2	7.4	
Czech Republic	44.1	7.4	⇩	62.9	3.9		60.9	5.0	

generation immigrant students, six of whom are in high-resourced schools, 17 in medium-resourced schools, and five in low-resourced schools.

Consequently, although the differences are statistically significant, they are likely to be an artifact of the data. Also for Morocco, we see a difference of 127 score points between students in well-resourced schools and medium-resourced schools. But although there are 169 first-generation immigrant students in the Moroccan data, only five of them are in high-resourced schools. We can also regard this difference as an artifact.

For Australia, Bahrain, and El Salvador, there are statistically significant differences of around 60 score points, and the statistics for these three countries are based on a good number of observations. In Australia there are 258 first-generation immigrant students in high-resourced schools and 152 in medium-resourced schools. In Bahrain there are 136 first-generation immigrant students in high-resourced schools and 430 in medium-resourced schools. In El Salvador there are 30 first-generation immigrant students in high-resourced schools and 108 in medium-resourced schools – which is probably too little a sample to provide reliable data.

The differences in the Canadian provinces cannot be attributed to a small number of observations. In Ontario, there are 154 observations in the high category and 204 in the medium category; in Quebec there are 162 and 157, respectively. Israel, Lebanon, Malta, Oman, Palestine, and the United States also have a good number

Table 4C.9 Achievement of first-generation immigrant students attending well-resourced schools

Country	High Math ach	SE		Medium Math ach	SE	Low Math ach	SE	
Thailand	513.0	49.5	⇑	337.6	30.6	380.5	48.5	
Morocco	447.3	41.5	⇑	320.0	12.2	315.7	8.3	
Botswana	424.1	35.8		356.3	14.8	296.4	13.6	⇓
Bahrain	424.6	8.2	⇑	360.2	5.0	385.5	16.3	
El Salvador	352.8	26.2	⇑	291.6	13.0	284.2	16.4	
Hungary	493.3	20.2		435.4	34.1	419.3	28.9	
Mongolia	444.7	34.4		387.0	7.6	380.5	8.5	
Australia	520.0	9.8	⇑	463.0	11.6	465.2	19.7	
Israel	469.6	11.7	⇑	424.0	10.4	458.2	21.2	
Turkey	453.9	32.1		410.1	21.7	353.8	46.8	
Oman	348.1	18.9	⇑	305.4	7.6	322.1	10.7	
Canada (Quebec)	535.4	15.0	⇑	495.3	8.4	571.5	15.9	⇑
Jordan	415.6	22.2		376.5	10.1	347.5	21.2	
Malaysia	447.2	7.9	⇑	409.3	13.4	466.3	24.8	⇑
Lebanon	444.5	10.5	⇑	408.7	5.3	350.1	39.2	
Malta	447.7	8.7	⇑	412.6	10.2	426.2	17.0	
Palestinian National Authority	361.3	14.5	⇑	327.3	8.1	299.3	17.5	
Tunisia	433.1	25.8		402.4	7.7	408.1	11.8	
Ukraine	422.6	51.1		393.1	9.0	392.2	14.6	
Canada (Ontario)	548.2	10.3	⇑	520.2	7.6	584.0	5.2	⇑
Russian Federation	521.0	14.4		495.7	10.5	443.5	82.3	
United States	479.7	8.8	⇑	455.5	7.1	434.7	14.0	
Indonesia	383.9	21.3		359.9	7.6	334.5	11.7	
United States (Massachusetts)	511.0	16.6		488.8	13.1	446.5	19.5	
Bulgaria	421.4	13.9		403.1	12.6	416.6	16.5	
Japan	548.2	24.5		532.1	30.8	NA		
Saudi Arabia	315.2	14.9		302.8	7.3	265.6	10.9	⇓
Bosnia and Herzegovina	462.9	23.0		451.3	4.7	454.4	14.2	
Norway	453.9	9.8		442.6	5.9	470.7	13.6	
Korea, Republic of	645.5	44.8		634.3	38.8	498.0	24.5	⇓
Italy	456.7	9.7		449.0	7.4	453.2	15.3	
Canada (British Columbia)	537.4	10.4		530.4	13.8	577.7	108.7	
Hong Kong, SAR	552.8	10.1		546.0	15.0	NA		
Egypt	355.3	12.0		349.5	6.3	345.8	26.6	
Kuwait	332.8	18.4		327.3	7.3	329.3	12.2	
England	497.1	19.7		493.3	13.9	496.1	17.2	
Chinese Taipei	498.3	17.2		495.6	13.2	481.8	31.6	
United Arab Emirates (Dubai)	493.1	4.6		490.6	7.0	431.0	16.1	⇓
Qatar	297.7	6.2		295.9	3.4	291.5	11.7	
Cyprus	421.8	24.8		421.4	5.9	456.7	14.8	⇑

(continued)

Table 4C.9 (continued)

Country	High Math ach	SE		Medium Math ach	SE	Low Math ach	SE	
Lithuania	443.2	12.4		445.6	10.6	453.7	18.7	
Syria, Arab Republic of	359.9	15.4		363.8	6.8	348.6	28.5	
Scotland	455.1	12.6		459.1	19.1	345.4	17.4	⇩
Sweden	455.5	9.0		461.0	7.7	485.3	12.2	
Slovenia	444.9	9.7		451.5	10.4	NA		
Iran, Islamic Republic of	367.6	49.0		374.4	23.2	381.5	71.4	
Romania	380.2	22.9		387.1	13.7	306.9	87.3	
Czech Republic	483.1	13.2		490.4	9.2	NA		
Singapore	619.5	6.2		635.9	15.6	NA		
Colombia	319.5	22.2		339.3	9.8	295.8	13.2	⇩
Georgia	354.1	29.0		374.7	12.1	361.2	43.0	
Spain (Basque Country)	443.8	9.1		469.2	15.5	350.1	83.5	
Serbia	437.2	34.7		462.6	9.1	441.5	26.4	
United States (Minnesota)	471.3	15.2		502.9	10.5	499.6	18.9	
Ghana	230.4	19.5		269.9	5.7	262.8	14.6	
Armenia	453.9	11.1	⇩	519.3	15.6	484.3	20.6	

of observations to make it unlikely that the calculated differences are only an artifact of the data. Anyhow, this does not mean that there is a causal effect between the school resources and the mathematics achievement of the students, but this is something that should be investigated further.

An interesting case is Armenia where the students in the high-resourced schools score statistically significantly lower than the students in the medium-resourced schools with a difference of 65 score points. This statistic is based on 80 observations in the high-resourced category and 367 observations in the medium-resourced category. This also requires some further investigations. Is it a special group of first-generation immigrant students in Armenia that attends high-resourced school and performs relatively poor? Or might there already be some policies in place that give better school resources to schools with relatively poor-achieving immigrant students?

When I examine the differences between the medium-resourced schools and the low-resourced schools for first-generation immigrant students, there are also some interesting differences. In Korea first-generation immigrant students in the low-resourced schools achieved 136 score points below their peers in medium-resourced schools in mathematics, or rather the student in the low-resourced school achieved below the peers in the medium-resourced school since there is a total of 20 first-generation immigrant students and only one of them in low-resourced schools.

Consequently this result is anything but representative for the country. In Scotland the difference is 114 score points, but there are only five students in the low-resourced schools. The difference of 119 points in Spain is not statistically

Table 4C.10 Achievement of first-generation immigrant students attending well-resourced schools

Country	High			Medium		Low		
	Math ach	SE		Math ach	SE	Math ach	SE	
Indonesia	544.9	28.2	⇑	377.2	25.3	298.2	20.6	⇓
Turkey	541.6	53.8	⇑	405.2	17.4	315.4	29.2	⇓
Morocco	475.7	22.3	⇑	363.6	15.1	352.5	13.3	
Korea, Republic of	586.4	16.3	⇑	478.5	20.7	672.9	48.0	⇑
Thailand	510.7	35.7	⇑	404.8	20.4	381.1	32.1	
Botswana	413.6	29.5	⇑	337.0	6.3	333.3	8.6	
El Salvador	403.0	20.9	⇑	329.0	9.5	343.0	11.9	
Lebanon	482.5	10.8	⇑	437.5	9.6	428.2	34.0	
Saudi Arabia	392.5	24.2		348.1	6.3	334.0	12.3	
Australia	518.5	10.0	⇑	476.3	8.5	484.7	36.9	
Japan	577.7	18.3		537.3	20.2	NA		
Canada (Quebec)	548.5	12.2	⇑	509.4	8.4	557.5	15.9	⇑
Hungary	545.2	21.4		510.9	13.1	548.2	8.6	⇑
Palestinian National Authority	395.1	16.3		362.7	9.8	354.7	13.3	
Tunisia	422.7	35.8		390.7	7.4	401.7	14.3	
United States (Massachusetts)	547.8	14.8		517.0	12.7	605.1	9.9	⇑
Egypt	385.2	20.1		357.0	10.6	333.6	23.3	
Malaysia	467.3	14.1		440.0	15.9	449.9	22.5	
Syria, Arab Republic of	408.5	20.7		381.3	8.4	410.9	16.4	
United Arab Emirates (Dubai)	474.1	5.5	⇑	447.4	7.5	412.4	36.2	
Oman	389.7	13.0		363.0	6.6	355.7	13.7	
Jordan	469.1	13.9		444.1	6.0	469.6	21.9	
Ukraine	499.9	16.5		475.3	5.9	480.8	13.2	
Cyprus	488.4	8.8	⇑	464.6	5.8	480.9	20.4	
Bosnia and Herzegovina	484.3	17.6		461.4	9.2	446.3	19.2	
Colombia	372.9	58.5		351.8	12.5	309.2	17.3	⇓
Serbia	513.5	15.9		493.6	5.7	480.3	12.9	
Malta	499.9	5.5	⇑	481.3	5.5	502.8	22.0	
Qatar	344.8	5.6	⇑	326.7	3.3	320.8	9.5	
Scotland	506.6	9.4		489.3	10.5	413.5	28.2	⇓
Canada (Ontario)	531.4	6.7		514.8	5.3	576.7	7.9	⇑
Iran, Islamic Republic of	387.3	32.9		374.9	15.2	334.4	23.4	
Bahrain	416.1	9.0		403.8	4.2	389.1	12.5	
Singapore	598.0	6.1		585.7	14.2	NA		
Russian Federation	511.2	10.3		501.1	8.8	491.9	24.7	
Georgia	374.2	32.1		365.3	17.5	340.9	31.9	
Norway	467.7	7.0		460.1	4.9	503.5	19.9	⇑
Italy	485.5	12.1		479.1	7.1	481.6	16.3	
Kuwait	366.3	12.1		362.5	5.5	372.4	19.7	

(continued)

Table 4C.10 (continued)

	High			Medium			Low		
Country	Math ach	SE		Math ach	SE		Math ach	SE	
United States	500.0	7.4		497.9	6.7		465.5	15.2	
Czech Republic	492.2	7.8		490.5	7.4		NA		
Israel	484.3	9.6		484.2	8.9		493.8	22.6	
Spain (Basque Country)	495.7	10.4		497.4	14.0		386.7	15.5	⇩
Canada (British Columbia)	517.0	4.8		519.0	5.0		561.1	42.0	
United States (Minnesota)	510.6	14.3		513.9	7.0		522.3	7.1	
Mongolia	394.3	15.2		400.0	5.8		386.4	7.1	
Slovenia	485.7	4.8		491.8	6.2		NA		
England	524.3	8.9		534.5	8.5		474.5	28.0	⇩
Hong Kong, SAR	577.8	8.0		588.2	10.0		NA		
Bulgaria	454.4	29.6		466.0	24.5		528.7	28.4	
Sweden	475.1	5.5	⇩	491.5	4.4		392.0	20.4	⇩
Lithuania	494.2	11.7		511.4	8.3		553.5	18.0	⇧
Chinese Taipei	580.6	20.7		605.3	16.0		530.0	60.0	
Ghana	272.6	29.7		309.6	10.9		278.0	20.2	
Armenia	465.9	22.4		513.8	13.3		508.3	15.6	
Romania	341.5	50.3		400.3	31.4		309.7	101.8	

significant due to the huge standard error for the achievement in the low-resourced schools, with only ten students in this category. Also in Botswana and Dubai, there are statistically significant differences of around 60 score points, but in Botswana there are 56 first-generation immigrant students in the low-resourced school category and only 23 in Dubai. In Colombia the 44-point difference is based on 79 observations in the medium-resourced category and 91 in the low-resourced category. The difference of 37 score points in Saudi Arabia is based on 526 observations in the medium-resourced category and 106 in the low-resourced category.

Interestingly there are a couple of participants where first-generation immigrant students in lower-resourced schools scored statistically significantly better than their peers in medium-resourced schools. This is the case for the Canadian provinces Ontario and Quebec, as well as Malaysia and Cyprus. But again in Ontario there are only 12 first-generation immigrant students in the low-resourced school category, two in Quebec, 20 in Malaysia, and 34 in Cyprus. Therefore, none of these results can be regarded as representative for the country.

This leaves us with the following countries where there is a higher mathematics achievement for first-generation immigrant students attending better-resourced schools: Australia, Bahrain, El Salvador, Ontario, Quebec, Israel, Lebanon, Malta, Oman, Palestine, Saudi Arabia, and the United States. Armenia is a special case where a negative relationship between school resources and mathematics achievement could be observed. All these cases should be examined further.

For second-generation immigrant students, the results are as follows. In 14 countries there is a statistically significant difference in mathematics achievement between the high-resourced and the medium-resourced schools. Countries with more than 100 score points that reveal statistically significant differences between these two groups are Indonesia (168), Turkey (137), Morocco (112), Korea (108), and Thailand (106).

There is only one country (Sweden with a difference of 16 score points) where the mathematics achievement is statistically significantly higher for second-generation immigrant students in the medium-resourced schools than in the high-resourced schools. However, one must again point out that these statistics are based on a small number of observations in some countries; in Morocco, there is even just one observation in the high-resourced category and two in Indonesia or Korea. This leaves us with the following eight countries where there is a statistically significant difference based on a fair number of observations: Australia, Canada Quebec, Cyprus (with only 61 observations in the high category), Lebanon, Malta, Qatar, Sweden, and Dubai.

For second-generation immigrant students, the results are somewhat more ambiguous when comparing students in medium- and in low-resourced schools. There are eight countries with students in medium-resourced school scoring statistically significantly better than students in low-resourced schools; on the other hand, there are seven countries where students in low-resourced schools score statistical significantly higher.

The highest statistically significant differences in favor of second-generation immigrant students in medium-resourced schools can be found in the Basque region of Spain (111), Sweden (100), Turkey (90), and Indonesia (80). On the opposite end, the participants with the highest statistically significant difference favoring the students in low-resourced schools are Korea (194), the state of Massachusetts (88), and the Canadian province Ontario (62). As stated before, none of the countries with statistically significant differences in either direction has a sufficient number of observations in the low category that makes the data trustworthy to be representative for the country.

As for the low category, the countries with the highest number of observations are Colombia with 36 and England with 27 observations. In other countries or participating systems, there are down to one or two observations in this category, for example, in Hungary, Korea, the Basque region of Spain, Sweden, or Turkey. I will therefore not investigate these differences any further.

As already discussed in Chap. 2, the students attending better resourced school might be the same students that come from more affluent homes. The TIMSS student background questionnaire includes a question about the number of books at the home. The data for this variable was already analyzed in Chap. 4B with respect to the different immigrant student groups. Table 4C.11 shows the average number of books at home for first-generation immigrant students in high-resourced, medium-resourced, and low-resourced schools. The table contains only those countries that have shown statistically significant achievement differences that are based on a sufficient number of observations.

Table 4C.11 Average number of books for first-generation immigrant students by school resources for selected countries

	High			Medium			Low	
Country	Number of books	SE		Number of books	SE		Number of books	SE
El Salvador	69	18.4		35	5.3		25	10.6
Canada (Ontario)	115	11.4	⇧	84	6.0		129	16.4
Armenia	82	13.4		56	6.8		100	15.0
Australia	99	8.0		76	10.0		74	24.6
United States	72	5.7	⇧	54	4.9		42	11.0
Bahrain	83	7.0	⇧	67	4.4		74	11.0
Canada (Quebec)	79	11.1		69	5.8		78	0.0
Lebanon	66	6.1		56	4.6		150	116.3
Palestinian National Authority	60	7.0		52	3.5		46	6.5
Israel	85	9.1		82	6.4		74	26.1
Oman	55	7.0		56	4.2		57	6.8
Saudi Arabia	57	12.3		60	5.5	⇧	36	4.0
Malta	93	6.4		101	6.3		109	16.7

We can see that there are quite some differences in the number of books – as a predictor for the socioeconomic background of the students – and the attendance of high-, medium-, or low-resourced schools. Only in few of the countries the differences are statistically significant.

In Bahrain, Saudi Arabia, and the United States, the socioeconomic background of first-generation immigrant students seems to have a highly significant positive relation to the school resources of the schools they attend. In the Canadian province of Ontario, the picture is different. The students in medium-resourced school have the lowest socioeconomic background – statistically significantly lower than students in high-resourced schools.

The highest socioeconomic background of the students, however, is found in the low-resourced schools – and is statistically significantly higher than for the students in the medium-resourced schools. Not surprisingly, in Armenia the socioeconomic background of the students in the low-resourced schools is the highest, which is in line with the results reported in Chap. 4B regarding the socioeconomic background measured by the number of books at home and their mathematics achievement.

To get an idea of the impact of more affluent homes and better resourced schools, two regression models were calculated. The first regression model regresses the mathematics achievement on the number of books at home; for the second model, the mathematics achievement was regressed on the number of books at home and the well-resourced schools together. Again, this analysis was only carried out for the countries where a statistically significant difference of the mathematics achievement for differently resourced schools was found and the results based on a sufficient number of observations. For a comparison, the results of these regression analyses are presented for first-generation immigrant students and native students in Table 4C.12.

Table 4C.12 Percent of variance of mathematics achievement explained by different regression models

Country	First generation immigrant			Native			Second generation immigrant		
	Books only (%)	Difference (%)	res + books (%)	Books only (%)	Difference (%)	res + books (%)	Books only (%)	Difference (%)	res + books (%)
Lebanon	2.7	7.7	10.3	4.3	3.6	7.9	2.88	8.59	11.47
Australia	17.6	6.0	23.6	13.3	2.3	15.6	12.35	4.67	17.02
Canada (Quebec)	10.5	5.2	15.7	10.6	4.5	15.2	5.59	3.48	9.08
El Salvador	3.6	4.0	7.6	5.2	2.5	7.7	5.09	1.32	6.41
Bahrain	3.0	3.7	6.7	3.9	1.2	5.1	3.59	0.60	4.19
Palestinian National Authority	0.5	2.6	3.1	2.6	1.1	3.7	4.99	1.80	6.79
Saudi Arabia	2.8	2.0	4.8	4.5	0.1	4.7	2.80	0.57	3.36
Malta	16.9	1.9	18.8	11.1	0.3	11.4	6.13	0.45	6.58
Israel	3.6	1.7	5.4	6.2	1.7	7.9	5.23	-0.18	5.05
Armenia	1.6	1.3	2.8	2.4	0.1	2.5	4.71	2.06	6.77
United States	11.2	1.0	12.2	13.4	-0.6	12.8	13.91	0.60	14.52
Oman	2.1	0.3	2.4	4.8	0.1	4.9	7.46	1.19	8.65
Canada (Ontario)	8.6	0.0	8.6	11.3	-2.8	8.5	7.33	0.02	7.34

A comparison of the variance explained by the models reveals that in few countries does the inclusion of well-resourced schools increase the explained variance substantially. Most prominent are the results for Lebanon where the number of books at home explained only 2.7 % of the variance of the mathematics achievement, whereas an additional 7.7 % of the variance of achievement could be explained by the indicator of well-resourced schools. Interestingly, for Lebanon the results for native students are quite different, and the bigger part of the achievement variance is explained by the books at home compared to the school resources. This might indicate that the school resources are especially relevant for immigrant students in Lebanon. Australia and the Canadian province of Quebec are other examples where the school resources explain more than 5 % of the variance of the mathematics achievement of first-generation immigrant students.

I conclude that overall immigrant students attend well-resourced schools to the same degree as native students. Yet, there are some countries where there are less first-generation immigrant students to be found in well-resourced schools, such as the Czech Republic, Qatar, Cyprus, Iran, and Malta. In Malta and Qatar, this holds also true for second-generation immigrant students. I also found that the resourcing is positively related to the students' achievement in some countries. From the literature, we know that students from more affluent homes attend better resourced schools. This was also found for the first-generation immigrant data in this study when measuring affluence by the number of books at home. I could, however, show that beyond this effect there is an additional effect explaining the better achievement in TIMSS mathematics of first-generation immigrant students who attend better resourced schools.

Table 4C.13 Percent of students in schools with a school climate rated highly positive by the principal by immigration status for selected countries

Country	First generation immigrant			Native			Second generation immigrant		
	Percent	SE		Percent	SE		Percent	SE	
United Arab Emirates (Dubai)	59	1.9	⇧	37	2.6		59	3.4	⇧
England	37	6.5		28	3.8		45	6.5	⇧
Qatar	26	0.8	⇧	22	0.5		22	0.7	
Bahrain	18	1.7		17	0.4		20	1.6	⇧
Cyprus	9	1.1		10	0.3		14	1.4	⇧
Malta	12	1.5	⇩	22	0.4		22	1.9	
Spain (Basque Country)	11	4.2	⇩	24	5.1		22	7.4	

School Climate

The next thing to examine is the school climate, how it might differ for immigrant and native students, and how it relates to achievement. As discussed in Chap. 2, the school climate can have a big impact on student achievement, and especially for students at risk, a positive school climate can influence the achievement positively. In TIMSS the school climate was assessed at teacher and at school level. For teachers as well as for principals, the TIMSS data includes an index on school climate, indicating high, medium, and low school climate.

Table 4C.13 shows the percentages of students in schools where the principal rated the school climate in the highest positive category. Only those countries are included where the percentage for at least one of the immigrant student groups is statistically significant different from the percentage for native students. The percentages of students in the medium and low categories are not shown, either since the information is mostly redundant to what is displayed here.

For Malta and the Basque region of Spain, the percentage of students in the highest category of school climate is statistically significantly lower for first-generation immigrant students than for native students. For Qatar and Dubai, the opposite is the case and there is a statistically significantly higher percentage of students among first-generation immigrant students attending schools where the principal indicated the school climate as very positive. A higher percentage of second-generation immigrant students is attending schools for which the principal indicated a highly positive school climate in Bahrain, Cyprus, England, and Dubai.

The next consecutive question is then: Does the difference of the school climate make a difference to the students' achievement? In general this question has already been answered in Chap. 2 – also for the results in TIMSS – but not with respect to immigrant students. Calculating the mathematics achievement of native students and the two immigrant student groups separately for high, medium, and low school climate, the result is mainly the same as reported in the TIMSS international report (Mullis et al., 2008, p. 356).

The students in schools whose school climate was rated highest by the principal achieved better than the ones in the medium category, who achieved better than the students attending schools where the principal rated the school climate lowest. Among the 56 analyzed countries (and benchmark participants), there are 34 where native students in schools with a high rating achieved statistically significantly better in mathematics than native students in the middle category. Moreover, there are 27 participants where native students achieved statistically better than native students in the low category.

There is one country (Qatar) where native students in the middle category achieved statistically significantly better than native students in the high category. What concerns first-generation immigrant students are that there are 19 countries where those in school with a high rating achieved statistically significantly better in mathematics than those in the middle category and there are 14 countries where this group achieved statistically significantly better than the group in the low category.

This is similar for second-generation immigrant students: in 19 countries, the students attending schools with a high rating achieved statistically significantly better in mathematics than those in the middle category; and there are 15 participating countries where the students achieved statistically significantly better than those in the low category.

The higher number of statistically significant differences for native students is by no means an indicator that there are more or higher differences for native students than for immigrant students but rather an effect of the smaller sample size that impacted the standard errors of the means. Indeed, the average difference for all countries between the achievement of students in the high and the medium categories is 25.0 score points for native students, 33.8 for first-generation immigrant students, and 35.0 for second-generation immigrant students. The achievement difference between students in the middle and the low category is 22.3 score points for native students, 26.3 for first-generation immigrant students, and 23.2 for second-generation immigrant students. Thus we can conclude that the school climate tends to make a difference in terms of mathematics achievement for all groups of students.

Since I have observed differences in the percentages of students in the three categories of school climate between native students and the two immigrant groups in seven countries, the results for these seven countries are presented in Table 4C.14. This table shows the mathematics achievement of immigrant students attending schools in the three categories of school climate as indicated by the principal.

As stated above, the same set of questions was also administered to the mathematics teachers of the students and the same indicator was calculated. Of course their perspective is a little bit different and shows statistically significant differences between immigrants and native students for a slightly different set of countries. Table 4C.15 shows the percentage of students where the school climate indicator coded to low for all groups of students according to the teachers' responses. Only countries are displayed where the percentages of native students differ statistically significant from at least one of the immigrant student groups.

Table 4C.15 shows that in Bahrain, Bulgaria, Lebanon, the Basque region of Spain, and in the United States, the percentage of first-generation immigrant students in the lowest category of school climate is significantly higher than of native students. In Qatar the opposite is the case and the percentage of first-generation immigrant students in the lowest category of school climate is significantly lower than of native students.

The highest difference can be observed for Bulgaria where 39.8 % of native students, as opposed to 64.7 % of first-generation immigrant students, attend schools where the mathematics teachers rated the school climate as low. For second-generation immigrant students, only the United States shows a statistically significant difference of the percentages with statistically significantly more second-generation immigrant students in schools where the mathematics teacher rated the school climate as low.

A clear pattern can be observed with regard to mathematics achievement in that the highest achievement mostly occurs in countries where students are attending

Table 4C.14 Mathematics achievement of students in schools with different levels of school climate as indicated by the principal by immigrant group

Country	Native High Math ach	SE		Medium Math ach	SE		Low Math ach	SE	
Bahrain	423.8	5.1	⇧	400.6	2.0		376.2	7.3	⇩
Cyprus	464.5	5.1		472.9	2.0		462.7	4.4	⇩
England	535.6	9.1	⇧	510.1	6.5		454.6	25.3	⇩
Malta	531.6	2.4	⇧	506.2	1.7		396.8	3.2	⇩
Qatar	296.0	3.8	⇩	309.6	2.0		277.9	4.8	⇩
Spain (Basque Country)	526.7	4.6	⇧	502.0	3.1		468.7	8.2	⇩
United Arab Emirates (Dubai)	423.9	11.8	⇧	382.5	8.5		394.2	20.7	

Country	First generation immigrant High Math ach	SE		Medium Math ach	SE		Low Math ach	SE	
Bahrain	411.4	7.5	⇧	372.1	4.5		341.7	13.5	⇩
Cyprus	412.8	25.1		424.3	6.3		443.4	13.2	
England	520.4	20.2		491.3	11.1		370.8	22.7	⇩
Malta	476.0	11.1		462.6	7.2		346.0	9.0	⇩
Qatar	298.9	4.4		293.3	3.5		277.3	9.3	
Spain (Basque Country)	480.4	40.4		444.0	11.2		445.3	12.4	
United Arab Emirates (Dubai)	501.5	5.5	⇧	478.6	5.3		394.7	12.0	⇩

Country	Second generation immigrant High Math ach	SE		Medium Math ach	SE		Low Math ach	SE	
Bahrain	444.0	7.7	⇧	399.8	4.0		352.4	12.8	⇩
Cyprus	474.0	16.4		470.4	5.6		460.6	13.2	
England	547.4	8.6	⇧	514.5	9.2		420.1	73.9	
Malta	514.2	7.4		503.5	4.1		404.5	10.2	⇩
Qatar	329.7	5.3		332.8	2.9		307.9	7.2	⇩
Spain (Basque Country)	507.3	17.0		502.7	10.9		470.5	14.9	
United Arab Emirates (Dubai)	481.8	4.5	⇧	444.3	6.0		432.5	33.6	

Table 4C.15 Percent of students with a school climate rated negatively by the mathematics teacher by immigration status for selected countries

Country	First generation immigrant Percent	SE		Native Percent	SE		Second generation immigrant Percent	SE	
Spain (Basque Country)	53.5	8.7	⇧	18.4	3.4		27.0	5.9	
Bulgaria	64.7	4.5	⇧	39.8	3.5		44.3	8.2	
Bahrain	34.7	3.6	⇧	23.6	1.7		23.3	2.5	
United States	30.4	3.4	⇧	19.7	2.0		30.4	3.7	⇧
Lebanon	20.4	4.0	⇧	9.9	2.1		7.9	2.4	
Qatar	20.0	0.9	⇩	22.2	0.5		21.2	0.8	

schools where the mathematics teacher rated the school climate as high, whereas it is the lowest among those students attending schools where the mathematics teacher rated the school climate as low.

When looking exemplary at the TIMSS results from the United States, native students in schools rated high on the school climate index by the mathematics teacher have an average mathematics achievement of 537 score points. Native students score 509 points in the middle category and 493 points in the low category. First-generation immigrant students attending schools rated high on school climate by the mathematics teachers have an average mathematics score of 505 points, the ones in the middle category of 457, and the ones in the low category of 426 points. The same pattern occurs for second-generation immigrant students, and we can observe an average mathematics score in TIMSS of 524 for the students in the schools ranking high on the school climate scale, 495 for the ones in the middle category, and 458 for the ones in the low category.

For all three groups of students, the achievement differences between the high and the middle categories and between the middle and the low categories are statistically significant except for native students in the low (493) and middle (509) category. It is notable that the achievement gap between the students in the different school climate categories is larger for immigrant than for native students. This suggests that the school climate might be more important for immigrant students than for native students. Since the group of immigrant students in the United Stated can be regarded as students at risk (Baca, Bryan, & McKinney 1993; Shields & Behrman, 2004), this observation follows the argument of Freiberg that a more positive school climate can be a measure for improving the achievement of students at risk (Freiberg, 1998).

I conclude that in general there is not much difference between the percentages of immigrant and native students in terms of attending schools where the principal rated the school climate positive. But I found seven participants where there is a statistically significant difference of the percentage of either first- or second-generation immigrant students compared to native students attending schools with a positive school climate: Bahrain, Cyprus, Dubai, England, Malta, Qatar, and the Basque region of Spain. For Malta and Spain, I observed a higher percentage of native students than first-generation immigrant students attending schools rated positively in terms of the school climate. For the other five participants, the difference is to the advantage of one of the immigrant groups. A more positive school climate – as rated by the principal – is related to a higher mathematics achievement for native as well as for immigrant students.

When examining the teacher's perspective on the school climate, I find six participating systems where there is a statistically significant difference of the percentage of either first- or second-generation immigrant students compared to native students attending schools with a negative school climate and these are Bahrain, Bulgaria, Lebanon, the Basque region of Spain, Qatar, and the United States. In all countries but Qatar, there is a higher percentage of first-generation immigrant students attending schools where the teacher rated the climate negatively. In the United States, the percentage of second-generation immigrant students attending schools with a school climate rated negatively by the teacher is higher than for the percentage of native students. In Qatar there is a lower percentage of fist-generation immigrant students attending schools where the teacher rated the climate negatively. As for the principal rating, I observed a positive relation between positive school climate and higher student mathematics achievement.

School Safety

One particular aspect of the school climate is that students feel safe in the school. As discussed in Chap. 2, there are two indices calculated for school safety in TIMSS. I based one of these on teacher-level information and the other on student-level information. In the following, I will take a deeper look at the student-level index on student safety. This index was calculated by questions that ask the students if any of the following has happened to them (Fig. 4C.1):

The index has three values: high, medium, and low. Students were assigned the high value if all five statements were answered negatively.

Table 4C.16 shows the percentages for native, first-, and second-generation immigrant students in the highest category together with an indicator for the immigrant student groups if the percentage is statistically significantly different from the one for native students. In 38 out of 53 countries, the percentage of first-generation immigrant students that are in the high level of feeling safe at school is statistically

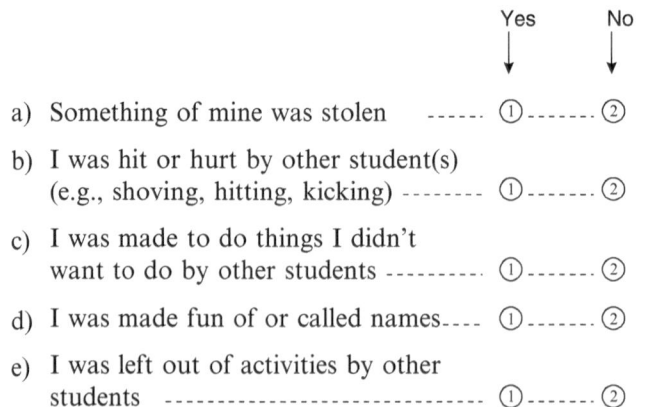

In school, did any of these things happen during the <u>last month?</u>

*Fill in **one** circle for each line*

                                                         Yes   No

a) Something of mine was stolen   ------ ① ------- ②

b) I was hit or hurt by other student(s) (e.g., shoving, hitting, kicking) -------- ① ------- ②

c) I was made to do things I didn't want to do by other students --------- ① ------- ②

d) I was made fun of or called names.---- ① ------- ②

e) I was left out of activities by other students ---------------------------- ① ------- ②

Fig. 4C.1 TIMSS 2007 student questionnaire, variables BS4GSTOL, BS4GHURT, BS4GMADE, BS4GMFUN, BS4GLEFT (SOURCE: TIMSS 2007 Assessment. Copyright © 2009 International Association for the Evaluation of Educational Achievement (IEA). Publisher: TIMSS & PIRLS International Study Center, Lynch School of Education, Boston College)

Table 4C.16 Percent of native and immigrant students that ranked feeling safe in school high

Country	First generation immigrant		Native students		Second generation immigrant		
	Percent	SE	Percent	SE	Percent	SE	
Canada (British Columbia)	53	3.3	48	1.3	48	1.6	
Australia	49	3.1	47	1.5	46	1.7	
Singapore	52	2.7	52	1.0	52	1.9	
Canada (Ontario)	45	3.2	46	1.8	50	2.6	
Hong Kong, SAR	51	1.8	52	1.5	51	1.5	
United Arab Emirates (Dubai)	47	2.4	50	2.6	44	3.0	
Botswana	8	1.9	10	0.6	6	1.1	⇩
Japan	62	8.2	65	1.1	51	6.7	⇩
Bosnia and Herzegovina	64	2.1	68	1.1	69	4.4	
Thailand	26	10.8	31	1.2	15	4.6	⇩
Canada (Quebec)	55	3.1	61	1.2	58	2.6	
Chinese Taipei	44	3.1	49	1.2	44	6.1	
Indonesia	31	2.2 ⇩	37	1.4	30	7.7	
Ghana	9	1.4 ⇩	15	1.0	15	2.6	
Russian Federation	66	3.2	72	1.1	67	2.3	⇩
Ukraine	63	3.0 ⇩	70	1.0	72	1.9	
England	51	3.3 ⇩	58	1.2	63	2.2	⇧
Turkey	43	7.2	50	1.5	50	6.1	
Colombia	33	6.1	41	1.5	28	5.2	⇩
Mongolia	39	2.4 ⇩	47	1.4	41	2.6	⇩
Bahrain	29	2.1 ⇩	38	1.1	37	2.1	
Cyprus	42	2.7 ⇩	52	1.0	46	2.6	
Tunisia	34	3.4 ⇩	44	1.2	30	3.8	⇩
Norway	56	3.1 ⇩	67	1.2	63	2.5	
Qatar	40	1.0 ⇩	51	0.8	47	1.3	⇩
Serbia	59	4.1 ⇩	70	1.2	67	2.4	
Israel	53	3.1 ⇩	64	1.4	58	2.1	⇩
Slovenia	43	4.5 ⇩	54	1.3	57	2.2	
Czech Republic	48	4.2 ⇩	59	1.2	61	3.2	
Sweden	64	2.7 ⇩	75	0.9	77	1.9	
Malta	40	2.5 ⇩	52	0.8	53	2.0	
El Salvador	41	4.4 ⇩	55	1.1	47	5.1	
Morocco	24	3.2 ⇩	38	1.5	32	5.0	
Malaysia	38	3.1 ⇩	52	1.5	46	6.7	
Kuwait	47	2.4 ⇩	62	1.2	54	2.4	⇩
Spain (Basque Country)	50	4.4 ⇩	65	1.7	58	5.0	
Iran, Islamic Republic of	34	7.4 ⇩	49	1.6	32	5.5	⇩
Jordan	40	2.5 ⇩	56	1.6	51	2.2	

(continued)

Table 4C.16 (continued)

Country	First generation immigrant			Native students		Second generation immigrant		
	Percent	SE		Percent	SE	Percent	SE	
Italy	48	4.0	⇩	64	1.2	59	3.3	
Palestinian National Authority	33	2.6	⇩	49	1.5	43	3.5	
Scotland	45	3.5	⇩	61	1.1	56	3.1	
Hungary	45	5.1	⇩	62	0.9	55	4.6	
Oman	34	2.1	⇩	51	1.4	40	2.6	⇩
Bulgaria	38	4.2	⇩	55	1.1	54	5.7	
Saudi Arabia	32	2.4	⇩	50	1.4	46	2.6	
Romania	30	4.1	⇩	49	1.1	23	7.7	⇩
Syria, Arab Republic of	35	1.7	⇩	55	1.3	46	3.2	⇩
Egypt	30	2.0	⇩	52	1.4	37	3.9	⇩
Lebanon	23	2.0	⇩	44	2.1	30	4.1	⇩
Lithuania	37	4.2	⇩	61	1.1	58	3.8	
Georgia	47	6.0	⇩	75	1.5	62	8.6	
Armenia	38	4.0	⇩	69	1.0	57	4.4	⇩
Korea, Republic of	19	7.7	⇩	52	1.3	33	13.1	

significantly lower than for native students. Only in British Columbia, Australia, and Singapore the percentage of first-generation immigrant students in this group is higher – but in none of the countries, this difference is statistically significant.

There are 17 countries with a statistically significantly lower percentage of second-generation immigrant students than native students in the high category. I have to point out once more that the statistics are based on very few cases in some countries and that the results for, e.g., Korea (with a 33 % difference for first-generation immigrant students) are not authoritative. This is different for the statistics for, e.g., Armenia that shows a difference of 32 % fewer first-generation immigrant students in the group of students classified as high on the student perception of school safety index which is based on 415 cases.

There is only one country with a higher percentage of students in the high school safety index among second-generation immigrant students compared to native students: England. The analysis of the percentage of native students and immigrant students in the category of low school safety reveals the same result with a significantly higher percentage of immigrant students than native students in this category.

This rather negative result – especially for the first-generation immigrant student population – should be seriously considered by policy makers and other stakeholders, such as teachers and principals, who should aim to determine measures for raising the perception of school safety for immigrant students.

According to the TIMSS international report, "There was a positive association between average mathematics achievement and students' perception of being safe at both fourth and eighth grades, with highest achievement among students at the high level of the index and lowest achievement among those at the low index level" (Mullis et al., 2008, p. 368).

When analyzing the mathematics achievement results for native and immigrant students separately for the three levels of the school safety index, the results reported in the TIMSS international report are confirmed for all three groups. The achievement differences are more often statistically significant for native students than for immigrant student populations although they are on average higher due to larger standard errors for the smaller-sized samples of immigrant students.

The results of the analysis are displayed exemplary for first-generation immigrant students. Table 4C.17 shows the mathematics achievement of immigrant students for the three levels of feeling safe at school. It is also indicated if the achievement for the high or low level of feeling safe is statistically significant different from the mathematics achievement of the students in the middle category of feeling safe at school.

The difference of more than 100 score points between the students in the high and medium category found for Japan – although statistically significant – is based on less than 50 students in total. Consequently, we should not rely on this result. But the results for the other countries that are marked as statistically significant different are based on a couple of hundred students each and can thus be trusted to be valid.

For example, in Botswana, the achievement difference between first-generation immigrant students in the high category of feeling safe at school and first-generation immigrant students in the medium category of feeling safe at school is more than 80 score points – a difference in learning of more than two school years. The first-generation immigrant students in the high category of feeling safe at school achieved even more than 120 score points in mathematics above first-generation immigrant students in the low category – a difference in learning of more than three school years.

For first-generation immigrant students, the difference between students in the high level of school safety and students in the medium level of school safety across all countries is 18 score points; between students in the medium level of school safety and the low level, the difference is 21 score points.

For native students, the results are not displayed in detail, but the average mathematics achievement difference between students in the high and medium levels of school safety across all countries is 9 score points; and between the students in the medium and low levels of school safety, it is 18 score points.

The results for second-generation immigrant students by country are not shown either. But for second-generation immigrant students, the difference between students in the high and medium levels of school safety across all countries is 18 score points, and between the students in the medium and low levels of school safety, it is 19 score points. Due to the standard error terms, I cannot say that the differences for immigrant students are higher than for native students, but I conclude that there is a clear relationship between the students feeling safe at school and their mathematics achievement, which also matches the previous research as laid down in Chap. 2.

Table 4C.17 Achievement differences for first-generation immigrant students with different levels of feeling safe in school

Country	High			Medium			Low		
	Math Ach	SE		Math Ach	SE		Math Ach	SE	
Japan	581.59	19.80	⇑	466.06	36.47		458.15	60.46	
Thailand	450.50	64.07		363.91	38.93		350.30	28.63	
Botswana	439.77	28.71	⇑	359.64	11.53		317.89	14.27	⇓
Serbia	473.62	8.92	⇑	430.32	17.15		434.19	29.93	
Georgia	394.26	16.95		354.48	13.98		373.33	28.54	
Lebanon	466.57	8.13	⇑	427.36	7.79		397.97	5.80	⇓
Jordan	411.19	9.14	⇑	373.05	8.97		376.92	24.90	
Oman	346.16	7.52	⇑	315.52	8.08		287.45	11.71	⇓
Qatar	323.45	5.26	⇑	294.46	4.61		259.86	5.07	⇓
Chinese Taipei	507.56	11.91		479.39	18.16		493.42	15.63	
Ghana	301.89	12.33		275.65	7.07		247.93	6.13	⇓
Egypt	375.60	5.86	⇑	349.41	6.42		336.68	8.03	
Sweden	472.02	6.31	⇑	447.49	10.27		417.99	10.61	⇓
United Arab Emirates (Dubai)	503.40	5.03	⇑	479.03	4.20		462.43	9.20	
Ukraine	412.81	10.80		389.15	12.81		357.44	23.95	
Turkey	433.93	32.79		410.86	23.58		317.71	18.71	⇓
Scotland	469.04	14.34		446.60	11.86		416.73	16.64	
Palestinian National Authority	359.00	11.36		336.76	8.76		296.59	12.15	⇓
Malaysia	444.93	7.00		423.18	9.56		412.96	15.80	
Mongolia	405.99	8.58		384.46	7.77		371.21	8.06	
Singapore	637.20	7.78		615.66	8.64		577.61	13.14	⇓
Kuwait	348.47	6.99		327.54	8.26		295.02	11.71	⇓
Canada (Quebec)	528.14	10.19		507.27	7.07		478.93	26.05	
Bahrain	399.72	8.60	⇑	379.24	5.84		349.48	9.75	⇓
Iran, Islamic Republic of	392.84	27.76		372.47	27.02		341.77	33.59	
Israel	461.24	8.72		441.69	9.85		400.34	18.32	⇓
Russian Federation	506.50	10.25		487.11	9.09		500.28	21.77	
England	509.74	13.02		491.00	12.96		459.57	19.09	
Bulgaria	421.61	11.41		405.48	15.06		394.35	18.27	
Syria, Arab Republic of	375.95	7.20		361.00	6.08		356.45	7.90	
Spain (Basque Country)	459.74	12.00		445.69	10.21		422.00	21.41	
Hong Kong, SAR	562.82	8.26		550.48	10.53		517.54	15.86	
Canada (Ontario)	542.80	5.96		531.72	7.90		503.98	14.76	
Romania	396.80	16.93		387.50	17.76		363.34	16.45	
Canada (British Columbia)	542.34	7.91		533.06	9.90		507.97	10.68	
Italy	459.30	7.60		450.72	8.36		399.76	27.30	

(continued)

Table 4C.17 (continued)

	High			Medium			Low		
Country	Math Ach	SE		Math Ach	SE		Math Ach	SE	
Colombia	327.31	13.73		318.92	10.98		308.60	11.44	
Bosnia and Herzegovina	460.81	5.55		452.56	6.17		409.40	11.95	⇩
Tunisia	409.07	8.77		402.63	8.73		410.43	12.56	
Malta	449.72	10.60		443.31	9.31		383.11	13.71	⇩
Slovenia	456.16	10.87		450.42	9.78		433.82	15.60	
Lithuania	456.58	17.55		452.03	11.93		424.44	13.58	
Norway	447.46	6.07		444.12	7.41		427.03	16.64	
Indonesia	358.32	9.01		355.65	7.10		336.68	10.22	
Australia	502.15	9.94		500.86	9.34		471.46	18.75	
Cyprus	438.86	8.57		437.85	8.03		402.33	12.83	⇩
Hungary	473.19	25.05		474.12	19.89		422.79	27.17	
Czech Republic	486.24	9.90		489.04	10.26		482.19	28.70	
Saudi Arabia	297.90	8.13		304.01	9.19		294.26	10.96	
El Salvador	299.01	18.09		305.26	12.29		303.60	13.75	
Morocco	317.15	18.13		324.59	12.10		327.22	13.36	
Algeria	368.66	48.01		383.21	23.98		NA		
Armenia	483.76	8.36		505.83	17.96		548.50	19.41	
Korea, Republic of	619.49	48.96		643.35	33.79		584.79	57.30	

I conclude that there is a difference in the percentage of students feeling safe at school between immigrant and native students. In the majority of countries, there is a statistically significantly lower percentage of first-generation immigrant students with a high level of feeling safe at school compared to native students. Also for second-generation immigrant students, in nearly one third of the participating countries, there is a statistically significantly lower percentage of students in the high category of feeling less safe in school. This is a matter of concern since for all three groups of students – native students and first- and second-generation immigrant students – there is a clear positive relation between feeling safe at school and mathematics achievement.

Summary

In this chapter, I found some differences in terms of where immigrant students are located within the countries. I concluded that in some countries immigrant students are often located in more urban areas. This is in particular true for second-generation immigrant students. But in other countries immigrant students are found more in rural areas, which creates different challenges for the education systems of the different countries. I noted some achievement differences between immigrants located in urban or rural areas for several countries. Some countries seem to be able to offer good opportunities to immigrant students mostly in large cities, but they face challenges in more rural areas.

I discovered a slightly higher percentage of first-generation immigrant students in schools with low school attendance. I also found school attendance to be associated with mathematics achievement and that the association is stronger for first-generation immigrant students in some countries.

I found immigrant students in general attending similar resourced schools as native students. However, in the Czech Republic, Qatar, Cyprus, Iran, and Malta, first-generation immigrant students attend less well-resourced schools. In Malta and Qatar, this is also true for second-generation immigrant students. I also found that the resourcing is positively related to the students' achievement in some countries – considering that students from more affluent homes attend better resourced schools.

And I discovered that in general immigrant students and native students are similarly attending schools where the principal rated the school climate positive. I determined that the school climate – as rated by the principal – is related to the mathematics achievement of native as well as of immigrant students. I found similar results when analyzing the teacher rating of the school climate; i.e., in general native students and immigrant students are attending schools with similar ratings of the school climate. There are, however, some countries that show significant differences. In general there is a positive relation between attending schools that are rated positively and students' mathematics achievement.

Lastly I found in this chapter major differences between immigrant and native students in terms of feeling safe in school. This is probably the most astonishing result. In the majority of countries, there is a statistically significantly lower percentage of first-generation immigrant students with a high level of feeling safe at school compared to native students. As for second-generation immigrant students in nearly one third of the participating countries, there is a statistically significant lower percentage of students in the high category of feeling less safe in school. For native as well as for first- and second-generation immigrant students, I determined a clear positive relation between their feeling safe in school and their mathematics achievement.

Especially for this last result, it is important to explore the underlying mechanisms and to find ways to improve the situation for immigrant students in terms of feeling safer in school. But this is clearly beyond the scope of this research and would probably require more qualitative approaches.

Chapter 4D
Class Level and Teacher Level Factors

Abstract In this chapter, class level factors and teacher-related factors that can influence student achievement will be evaluated. Small class sizes are assumed to have a positive impact on student achievement and the author will explore if this is supported by TIMSS data. Teacher-related factors became increasingly important in educational research and policy recommendations; opinions on the influence of teacher quality, however, vary widely. The impact of homework given by teachers will be examined, and differences in the consequent achievement of native and immigrant students will be examined in detail. To follow is an analysis of the distribution of immigrant students across classes in the countries that participated in TIMSS 2007, and the effects on student achievement, which prove to form quite an ambiguous picture. How many immigrant students are there in the classes, and does their presence have a negative effect on the performance of native students? Are immigrant students more affected by class level or teacher-related factors than native students? The answers can be found in this chapter.

Keywords Class size • Homework • Peer effects

In this chapter, I will evaluate the class level factors that can influence student achievement, such as class size, but I will also consider teacher-related factors. Teacher-related factors became increasingly important in educational research and policy recommendations. The OECD (2005, p. 23) stated: "The research indicates that raising teacher quality is perhaps the policy direction most likely to lead to substantial gains in school performance." On the other hand, Hattie (2008, p. 108) has a much more differentiated perspective on teacher effects and stated that the fact that "… teachers make a difference is misleading. Not all teachers are effective, not all teachers are experts, and not all teachers have powerful effects on students…."

Class Size

I will start examining the sizes of the classes attended by immigrant students compared to those attended by native students. As discussed in Chap. 2, class size is found to be related to student achievement in some studies, while other researchers

© Springer International Publishing Switzerland 2016

D. Hastedt, *Mathematics Achievement of Immigrant Students*,

DOI 10.1007/978-3-319-29311-0_7

contradict the assumption that small classes impact student achievement positively. TIMSS grouped grade eight students in three groups: small classes with 1–24 students, medium-sized classes with 25–32 students, and big classes with 33 students and more. Table 4D.1 shows the percentage of native students as well as first- and second-generation immigrant students in small classes.

The table indicates if the percentage of one of the immigrant student groups differs statistically significantly from the percentage of native students. Although we cannot identify statistically significant differences for most countries, there are differences for some. In Bahrain, Bulgaria, Malta, Qatar, Romania, and Scotland, there is a statistically significantly higher proportion of first-generation immigrant students in smaller classes than native students.

In the United States, the opposite is the case and there are statistically significantly more native students in small classes than first- – and also second- – generation immigrant students. Also in Jordan and Norway, there are statistically significantly more native students in small classes than second-generation immigrant students.

Although it is intuitively evident that there are better opportunities in smaller classes for supporting students that are facing language difficulties or other problems, the question remains if I can find higher achievement of immigrant students in small classes. Table 4D.2 shows the mathematics achievement of first-generation immigrant students in small-, medium-, and big-sized classes. But counterintuitively also for first-generation immigrant students, the same comparison as for all TIMSS students is true, namely, that first-generation immigrant students in small classes achieve on average less well than their immigrant peers in medium-sized classes. There are three countries with the opposite result: in Bahrain, Botswana, and Malaysia, first-generation immigrant students in small classes outperform their peers in medium-sized classes significantly.

However the curvilinear relationship that was found for all TIMSS students ("The complexity of this issue is evidenced in the TIMSS 2007 results showing a curvilinear relationship, on average, between class size and mathematics achievement at both the eighth and fourth grades" (Mullis et al., 2008, p. 273)) cannot be confirmed. In some countries, first-generation immigrant students in big classes perform less well than their peers in medium-sized classes, but in other countries, the opposite is the case.

While for native students the average achievement in medium-sized classes is 8 score points above the achievement of students in big classes, which is statistically significant, there is a difference of less than 1 score point between first-generation immigrant students in medium- and large-sized classes which is far beyond any statistical significance. This means that big classes have a lower association with achievement for first-generation immigrant students than for native students. This needs some further research.

Although the number of first-generation immigrant students in the TIMSS sample and their distribution in different types of communities and class sizes does not allow me to compare these groups in all countries, I can work with the results in

Table 4D.1 Percent of students in classes with 1–24 students by immigrant status

Country	First generation immigrant			Native		Second generation immigrant		
	Percent	SE		Percent	SE	Percent	SE	
United States (Minnesota)	47.3	11.4		29.8	5.7	35.4	6.6	
Bulgaria	73.0	5.4	⇧	57.7	3.5	49.2	7.5	
Romania	90.5	2.5	⇧	75.3	3.0	83.6	6.6	
Scotland	56.7	5.5	⇧	41.5	3.2	45.2	4.6	
United States (Massachusetts)	76.5	8.0		62.3	5.8	70.1	8.7	
Lithuania	44.5	6.1		33.5	3.2	39.9	6.2	
England	38.9	6.2		29.0	3.8	28.9	5.2	
Qatar	26.6	1.2	⇧	17.8	0.5	17.8	0.8	
Bosnia and Herzegovina	54.5	5.0		45.8	3.8	49.8	6.4	
Ukraine	44.6	6.5		36.9	3.2	27.9	4.1	
Malta	77.3	2.7	⇧	70.6	0.4	70.2	2.0	
Israel	9.2	4.5		2.8	0.6	6.3	2.1	
Spain (Basque Country)	73.1	6.2		67.5	2.8	66.5	5.9	
Colombia	16.6	5.6		12.3	2.4	24.1	8.0	
Italy	77.8	4.1		73.5	2.9	69.4	4.7	
El Salvador	38.7	6.0		34.6	3.7	37.0	5.7	
Indonesia	9.2	3.4		5.7	1.6	7.3	4.2	
Lebanon	42.0	5.3		38.6	4.8	35.7	5.2	
Hungary	74.3	5.9		71.2	3.4	80.1	4.6	
Bahrain	8.5	1.2	⇧	5.6	0.7	7.0	1.3	
Russian Federation	65.4	6.3		62.6	2.6	59.0	4.1	
Botswana	3.6	2.8		0.9	0.6	0.8	0.5	
Ghana	15.5	3.4		12.7	2.5	11.8	3.6	
Sweden	65.9	4.9		63.2	3.8	59.7	5.0	
Canada (Ontario)	39.1	8.8		36.4	3.8	34.9	5.1	
Hong Kong, SAR	12.2	2.3		9.8	2.6	8.5	1.6	
Armenia	42.1	5.7		39.9	4.0	38.3	6.6	
Canada (British Columbia)	32.6	5.3		31.0	4.7	28.0	3.9	
Thailand	12.5	9.6		11.0	2.4	21.5	12.2	
Palestinian National Authority	8.8	2.5		7.8	1.7	4.6	1.9	
Slovenia	95.3	1.3		94.4	1.0	94.7	1.5	
Tunisia	3.6	1.7		2.8	1.2	4.8	2.7	
Turkey	19.4	7.6		18.6	3.5	15.2	6.2	
Georgia	53.1	7.3		52.4	5.4	50.3	9.4	
Singapore	2.1	0.8		1.5	0.6	1.8	0.8	

(continued)

Table 4D.1 (continued)

Country	First generation immigrant		Native		Second generation immigrant		
	Percent	SE	Percent	SE	Percent	SE	
Chinese Taipei	4.2	2.1	4.2	1.8	6.1	3.6	
Oman	9.6	2.9	9.7	2.2	15.1	3.4	
Kuwait	12.1	3.6	12.5	3.4	9.7	2.9	
Mongolia	8.4	2.7	9.1	2.2	8.2	3.0	
Malaysia	0.4	0.4	1.4	0.8	0.9	0.7	
Morocco	5.7	3.3	6.8	2.7	5.0	2.2	
Egypt	3.3	1.4	4.5	1.6	5.6	2.6	
Cyprus	53.0	3.7	54.2	2.8	54.0	3.3	
Serbia	51.1	5.7	53.0	3.9	54.3	5.1	
Japan	7.5	5.5	9.7	2.2	15.7	7.0	
Syria, Arab Republic of	21.8	3.7	24.3	3.9	24.1	5.3	
United Arab Emirates (Dubai)	29.7	4.2	33.3	5.4	23.8	3.7	
Australia	27.8	4.1	31.4	3.4	26.9	2.9	
Saudi Arabia	25.7	4.0	29.4	4.0	23.2	3.8	
Korea, Republic of	NA	-	4.1	1.4	21.2	15.4	
Jordan	9.9	2.8	15.5	3.4	7.7	1.3	⇩
Iran, Islamic Republic of	28.8	12.3	35.5	3.2	28.3	4.9	
Czech Republic	41.9	7.7	49.7	4.3	43.2	5.6	
Canada (Quebec)	12.6	2.8	21.1	3.8	16.8	5.1	
Norway	37.7	4.5	49.0	3.9	36.4	4.3	⇩
United States	48.5	3.9 ⇩	61.7	2.5	44.2	3.6	⇩

Hong Kong. Hong Kong is a country where immigrant students in larger classes outperform immigrant students in smaller classes, and the number of immigrant students is quite large. Due to the structure of Hong Kong, there are no students from communities with less than 3,000 inhabitants, and I can distinguish only between students in communities with less or more than 500,000 inhabitants.

Table 4D.3 shows the percentages of first-generation immigrant students in Hong Kong in classes with 1–24 students, 25–40 students, and more than 40 students, and their mathematics achievement differentiated between the two abovementioned communities. One can see that in the larger communities, there is a higher percentage of first-generation immigrant students in larger classes with more than 40 students (35 % compared to 60 %). I also find that the mathematics achievement increases with class size from small- to medium-sized classes in both types of communities, but very differently so. Whereas the achievement difference between small- and medium-sized classes in communities with less than 500,000 inhabitants is only 34 score points, it is 129 score points in the communities with more than 500,000 inhabitants. Another difference is that there is no increase between medium and large classes in communities with more than 500,000 inhabitants.

Table 4D.2 Mathematics achievement of first-generation immigrant students attending different class sizes

Country	1–24			25–40		41 or more		
	Math ach	SE		Math ach	SE	Math ach	SE	
Botswana	475	42.3	⇧	355	13.6	331	19.0	
Bahrain	453	12.0	⇧	365	5.5	447	17.2	⇧
Malaysia	502	19.5	⇧	421	8.8	463	13.8	⇧
Iran, Islamic Republic of	400	34.7		366	21.3			
Armenia	527	20.5		494	10.7	437	22.3	⇩
Ghana	275	13.9		249	9.6	271	8.9	
Palestinian National Authority	348	15.8		324	10.2	330	13.0	
United Arab Emirates (Dubai)	500	10.6		478	5.7	352	20.7	⇩
Thailand	364	91.7		342	38.3	448	57.1	
Morocco	338	13.5		319	10.0	324	16.5	
Georgia	379	16.8		362	17.2	400	16.1	
Serbia	463	16.0		448	11.0			
Egypt	365	17.4		354	7.2	350	6.8	
Czech Republic	492	12.7		484	9.4			
United States	469	6.5		462	8.7	533	14.6	⇧
Syria, Arab Republic of	366	11.0		361	6.8	373	22.3	
Turkey	418	67.9		412	19.9	358	24.2	
Cyprus	431	7.3		427	8.2	341	28.5	⇩
Mongolia	389	20.8		386	6.4	430	17.7	⇧
Oman	317	25.7		316	6.7			
Saudi Arabia	294	12.2		298	9.3	288	22.2	
Israel	441	53.6		448	8.4	557	70.6	
Norway	441	7.5		450	5.9	418	13.7	⇩
Jordan	377	32.4		385	12.2	388	13.3	
Colombia	311	21.5		321	11.2	318	12.7	
Canada (Ontario)	527	12.5		536	9.1	543	7.6	
Italy	448	7.0		461	9.9			
Qatar	285	6.6		298	3.1	278	20.5	
Kuwait	326	18.1		340	7.1	438	17.1	⇧
Canada (Quebec)	498	9.3		513	9.4	506	14.2	
Bosnia and Herzegovina	443	6.4	⇩	463	5.3			
Hungary	454	25.6		476	18.9			
Canada (British Columbia)	516	9.9		542	10.6	487	42.9	
Indonesia	323	13.1	⇩	352	5.7	356	12.2	
Lebanon	404	7.9	⇩	434	9.5	399	31.3	
Sweden	450	7.2	⇩	481	9.5	490	17.7	
Ukraine	379	13.0		411	12.5	405	7.8	
Russian Federation	488	11.1	⇩	521	11.3			
El Salvador	275	14.6		308	14.8	311	19.7	

(continued)

Table 4D.2 (continued)

Country	1–24 Math ach	SE		25–40 Math ach	SE	41 or more Math ach	SE	
United States (Minnesota)	466	17.6		503	10.3			
Chinese Taipei	441	49.0		485	10.7	568	51.8	
Slovenia	445	7.7	⇩	490	17.5			
Japan	482	171.6		530	18.3	701	29.4	⇧
Bulgaria	402	10.3	⇩	451	16.8			
Malta	416	7.3	⇩	468	11.7			
Spain (Basque Country)	442	9.7	⇩	496	15.2			
Lithuania	413	15.3	⇩	468	9.6			
Hong Kong, SAR	483	19.3	⇩	542	13.5	587	11.7	⇧
United States (Massachusetts)	482	11.8	⇩	543	16.8	570	8.6	
Australia	455	12.9	⇩	519	9.5	452	17.8	⇩
Tunisia	331	34.9	⇩	405	6.5			
Singapore	549	44.9		628	7.2	609	11.7	
Scotland	413	13.7	⇩	500	11.0			
Romania	370	12.6	⇩	467	24.8			
England	429	19.2	⇩	535	10.5	446	100.8	
Korea, Republic of			⇩	646	34.3	625	25.8	

Table 4D.3 Percentages of first-generation immigrant students in Hong Kong in different mathematics class sizes for lager and smaller community sizes and their mathematics achievement

	Between 3.001 and 500.000					
	1-24	SE	25-40	SE	41 or more	SE
Percentage	16	3.3	49	6.8	35	6.0
Math score	483	19.8	517	14.3	568	18.4
	More than 500.000					
	1–24	SE	25–40	SE	41 or more	SE
Percentage	4	2.7	35	9.1	60	9.2
Math score	484	55.4	613	15.7	608	9.3

Table 4D.4 shows that the distribution of native and first-generation immigrant students in different class sizes is very similar for both community types. However, there are differences between native students and immigrant students in terms of achievement and not only in terms of overall achievement levels but also in terms of the differences between the groups. The most extreme difference can be observed for students in small classes in communities with more than 500,000 inhabitants. It is relatively much higher for native students – although the mean achievement is predicted with a big standards error for immigrant students and even more so for native students. These results show that there is also an effect of urban and rural communities linked to the class size effect.

Table 4D.4 Percentages of native students in Hong Kong in different mathematics class sizes for larger and smaller community sizes and their mathematics achievement

	Between 3.001 and 500.000					
	1–24	SE	25–40	SE	41 or more	SE
Percentage	13	3.5	53	6.6	35	5.6
Math score	514	24.6	537	11.3	600	10.2
	More than 500.000					
	1–24	SE	25–40	SE	41 or more	SE
Percentage	6	4.3	31	6.0	63	7.2
Math score	590	111.6	623	16.2	608	12.8

The achievement results for second-generation immigrant students across all countries are overall very similar to the ones for the first-generation immigrant students and will consequently not be displayed or further discussed here.

> I conclude that there are differences in various countries between the class sizes of classes attended by immigrant students and native students. But the differences go both ways. In some countries, immigrant students attend smaller classes, and in some countries, native students attend smaller classes. The relation between class sizes and mathematics achievement is also rather complex. Although in some countries immigrant students attending smaller classes perform better in mathematics, mostly the opposite is the case and I can only speculate why. I could find some associations of class sizes with community sizes, and as I have shown previously, the community sizes are also a factor for learning environments and in particular for the learning outcome – in this case, mathematics achievement. But beyond this, the data available does not give us much more insight.

Homework

The next thing to investigate is the homework assigned by the mathematics teacher. As seen in the literature research in Chap. 2, homework can be a factor in education. In TIMSS the mathematics teachers of the sampled students are administered a set of five questions about the mathematics homework that they assign to their students. Among them there are also questions about the frequency and expected time that it takes for an average student to finish them. In TIMSS an indicator was developed that reflects the teachers' emphasis on mathematics homework. "Students in the high category had teachers who reported giving relatively long homework assignments (more than 30 min) on a relative frequent basis (in about half of the lessons or more). Students in the low category had teachers who gave short assignments (less than 30 min) relatively infrequently (in about half the lessons or less). The medium level includes all other possible combinations of responses" (Mullis et al., 2008, p. 302).

Table 4D.5 shows the percentage of students in the low level of mathematics teachers' emphasis on homework for native students and the two groups of immigrant students. As can be seen in the table, there are only three participants where the percentage of first-generation immigrant students in this category is statistically significantly lower than the percentage of native students, namely, the two Canadian provinces British Columbia and Quebec and Singapore. For the third Canadian province, Ontario, there is not a statistically significant difference but slightly below the threshold. As stated above, these countries are also the countries where first-generation immigrant students perform relatively high compared to native students. One might hypothesize that there is a positive relationship between the immigrant students' higher achievement and the teacher's relatively higher emphasis on homework. But further research in this area is necessary to prove the relationship and even more its causal nature.

For second-generation immigrant students, the percentages appear to be smaller than for native students in Bulgaria, Qatar, and Sweden. The statistics for Bulgaria are based on 76 second-generation immigrant students, one of which is in the low category. Consequently, the statistics cannot be regarded as trustworthy.

More interesting are the results when comparing the relations of the assigned mathematics homework to achievement. In Table 4D.6 the mathematics achievement of native students is tabulated for the three groups of classes with different teacher emphasis on mathematics homework. Interestingly, there are several countries where the achievement differs statistically significantly between the different groups.

There are three countries where the mathematics achievement of students with mathematics teachers who put a high emphasis on mathematics homework was statistically significantly lower than that of students with mathematics teachers who put a medium emphasis on mathematics homework. But there are 12 countries where the mathematics achievement of students whose mathematics teachers put a high emphasis on mathematics homework was statistically significantly better than the achievement of students whose mathematics teacher put a medium emphasis on mathematics homework.

On average the students in the high category achieved seven score points above the students in the medium category across all countries. When focusing only on the countries where there is a statistically significant difference, the average difference between the students in the high category and the medium category amounts to even 32 score points. The achievement difference between students in the medium and low category is even more pronounced.

There is no country where students in the low category achieved statistically significantly better than students in the medium category. In 18 out of 55 countries, students whose mathematics teachers put a low emphasis on homework performed statistically significantly lower than those whose mathematics teachers put a medium emphasis on homework.

Table 4D.5 Percentage of students in mathematics classes with a low emphasis on homework

Country	First generation immigrant			Native		Second generation immigrant		
	Percent	SE		Percent	SE	Percent	SE	
Korea, Republic of	37.5	11.5		55.6	3.3	32.1	16.3	
Canada (British Columbia)	11.0	2.8	⇩	23.9	3.5	15.5	2.6	
Canada (Quebec)	14.8	4.6	⇩	27.4	4.2	16.2	4.4	
Canada (Ontario)	19.8	4.7		32.1	5.3	22.9	4.2	
United Arab Emirates (Dubai)	18.5	3.6		30.8	7.9	19.4	4.4	
Czech Republic	67.7	7.4		78.1	3.2	73.2	4.5	
Thailand	0.0	NA		9.2	2.2	8.1	4.3	
Sweden	56.8	4.8		65.6	3.0	53.4	4.8	⇩
Singapore	12.1	2.4	⇩	19.6	2.4	17.3	2.5	
Norway	11.4	3.0		18.7	3.3	12.0	3.1	
Iran, Islamic Republic of	6.9	7.0		13.2	2.8	14.7	4.8	
Bahrain	42.7	4.5		48.8	2.9	39.1	4.0	
Australia	46.9	5.6		51.8	4.4	43.7	4.2	
Syria, Arab Republic of	20.2	3.8		23.5	4.0	23.6	5.1	
England	57.4	6.6		60.3	4.2	47.7	6.3	
Ghana	19.6	4.0		21.9	3.6	17.9	4.3	
Colombia	14.2	4.8		16.0	3.2	20.6	5.6	
Italy	0.0	NA		1.5	0.8	0.5	0.5	
Armenia	9.3	3.4		10.3	2.5	10.2	3.2	
Jordan	29.4	5.5		29.8	4.2	20.7	4.2	
Cyprus	0.2	0.2		0.6	0.6	0.0	NA	
Romania	1.0	0.7		1.2	0.8	0.0	NA	
Serbia	27.2	6.0		27.4	3.9	28.3	4.8	
Ukraine	0.8	0.8		1.0	0.8	1.3	0.9	
Bosnia and Herzegovina	25.3	4.3		25.5	3.7	23.6	5.9	
Malaysia	12.4	3.9		11.6	2.3	7.8	2.8	
Israel	6.7	1.5		5.9	1.5	9.0	2.0	
Tunisia	7.0	3.7		6.1	2.0	5.7	3.3	
Morocco	14.0	4.4		13.2	2.7	17.0	3.8	
El Salvador	24.5	5.5		23.6	3.9	28.7	6.4	
Georgia	3.0	1.8		2.0	1.2	7.7	5.2	
Kuwait	83.3	4.1		81.7	3.8	84.6	4.3	
Lithuania	8.0	3.0		6.2	1.8	7.3	2.9	
Qatar	41.2	1.2		39.3	0.6	33.1	1.1	⇩
Malta	7.8	1.8		5.9	0.2	7.4	1.0	
Egypt	33.1	4.9		31.1	4.4	33.6	5.9	
United States	15.6	3.4		13.2	2.3	13.4	2.6	
Mongolia	8.1	3.6		4.6	1.6	4.7	2.3	
Lebanon	12.6	3.7		8.7	2.4	10.7	4.1	

(continued)

Table 4D.5 (continued)

Country	First generation immigrant			Native			Second generation immigrant		
	Percent	SE		Percent	SE		Percent	SE	
Hong Kong, SAR	19.7	4.7		15.3	3.7		16.1	3.6	
Botswana	13.9	3.7		9.2	2.5		10.6	2.9	
Turkey	32.7	8.1		27.6	3.3		42.3	9.1	
Slovenia	10.8	3.8		5.4	1.3		7.2	2.2	
Bulgaria	12.0	4.8		5.4	1.6		1.1	1.1	⇩
Scotland	60.5	5.8		53.8	3.7		60.4	4.8	
United States (Massachusetts)	13.3	5.7		6.6	2.2		13.1	5.0	
Indonesia	17.8	5.1		10.8	2.5		6.8	4.1	
Japan	65.4	10.4		58.4	3.8		71.0	6.3	
United States (Minnesota)	15.4	8.2		8.3	3.1		19.9	12.2	
Hungary	13.3	5.5		5.1	1.4		4.6	2.1	
Palestinian National Authority	35.7	5.7		25.9	3.5		27.0	5.1	
Spain (Basque Country)	20.4	7.8		10.5	2.6		16.0	5.9	
Oman	35.1	4.9		25.1	3.2		31.0	6.0	
Chinese Taipei	34.8	4.9		24.1	3.5		26.5	6.5	
Saudi Arabia	58.6	5.3		46.2	4.0		40.5	5.3	

Table 4D.6 Mathematics achievement of native students by mathematics teachers' emphasis on homework

Country	High			Medium		Low		
	Math ach	SE		Math ach	SE	Math ach	SE	
Czech Republic	575.7	28.3	⇧	505.0	7.7	502.1	3.1	
United States (Minnesota)	575.3	10.8	⇧	532.9	4.7	493.5	15.4	⇩
Bulgaria	502.7	8.7	⇧	461.5	6.3	460.3	9.8	
United States (Massachusetts)	585.5	9.6	⇧	547.2	5.3	508.1	11.1	⇩
England	556.1	11.5	⇧	521.3	10.5	500.0	6.3	
Israel	490.3	5.7	⇧	457.7	7.7	408.9	24.1	
Romania	476.6	4.8	⇧	445.4	9.1	407.8	13.0	⇩
United Arab Emirates (Dubai)	422.1	38.6		394.7	14.8	403.9	21.9	
United States	542.2	5.8	⇧	515.5	3.9	483.2	6.1	⇩
Malta	514.9	2.6	⇧	492.9	1.5	421.5	6.7	⇩
Scotland	535.6	14.5		513.7	6.1	468.5	5.3	⇩
Canada (British Columbia)	521.2	6.5	⇧	500.2	3.5	487.5	8.6	
Singapore	608.4	6.0	⇧	587.4	6.8	539.1	12.5	⇩
Korea, Republic of	608.6	7.6		591.6	5.8	597.5	4.0	
Ukraine	473.2	5.3		460.8	5.6	444.7	6.6	
Iran, Islamic Republic of	407.6	5.0		395.5	8.4	413.8	12.0	
Cyprus	479.0	3.9	⇧	467.4	2.2	463.5	9.4	

(continued)

Table 4D.6 (continued)

Country	High Math ach	SE		Medium Math ach	SE	Low Math ach	SE	
Morocco	397.7	8.1		386.1	4.6	381.2	9.0	
Thailand	448.7	7.7		437.8	9.0	439.7	14.0	
Georgia	421.5	8.0		411.4	7.9	408.9	37.8	
Ghana	329.6	8.6		319.6	7.9	323.5	7.4	
Botswana	375.7	4.4		366.1	3.7	357.0	6.7	
Hungary	527.7	14.2		518.8	3.7	484.7	17.7	
Mongolia	452.5	5.1		444.4	6.1	404.3	13.5	⇩
Syria, Arab Republic of	411.5	4.7		403.4	7.6	406.4	8.0	
Hong Kong, SAR	589.7	11.2		581.8	10.2	549.8	17.1	
Chinese Taipei	621.1	7.5		613.3	4.8	569.7	7.5	⇩
Russian Federation	519.1	6.1		511.8	4.4			
Turkey	433.9	9.6		428.7	8.8	434.9	10.4	
Armenia	499.9	3.6		495.1	5.4	499.8	12.8	
Malaysia	483.8	7.9		479.8	6.6	463.5	15.9	
Canada (Quebec)	543.6	12.6		540.4	5.1	509.8	5.3	⇩
Sweden	509.7	7.4		507.0	3.6	494.3	2.3	⇩
Oman	391.8	12.2		389.4	4.2	375.0	5.9	⇩
Slovenia	511.5	7.7		509.2	2.6	491.3	12.0	
Italy	482.9	3.4		481.6	5.3	392.6	25.5	⇩
Canada (Ontario)	517.0	5.3		515.8	4.7	506.6	11.5	
Colombia	387.1	4.5		387.9	7.6	371.5	10.2	
Spain (Basque Country)	505.2	6.0		506.1	3.6	499.8	7.8	
Jordan	429.2	13.1		432.8	6.0	420.1	9.0	
Egypt	424.9	8.6		428.7	4.7	424.7	7.5	
Norway	471.9	3.8		476.9	3.3	471.4	5.5	
Serbia	485.0	7.4		490.4	4.9	485.7	7.3	
Tunisia	420.1	3.5		426.7	3.3	425.0	10.3	
El Salvador	340.6	6.9		347.8	4.0	334.6	6.9	
Lithuania	502.8	5.6		511.8	2.6	484.1	6.0	⇩
Japan	563.3	7.7		576.0	4.6	569.7	3.9	
Indonesia	403.9	7.9		417.1	6.7	393.0	9.9	⇩
Bosnia and Herzegovina	449.4	9.7		463.5	4.0	449.8	5.9	
Lebanon	451.5	5.1		467.6	7.1	445.8	14.7	
Bahrain	388.1	5.7	⇩	405.3	2.8	397.9	3.2	
Palestinian National Authority	367.1	14.8		386.1	4.3	368.2	6.4	⇩
Saudi Arabia	316.5	17.7		338.2	4.0	331.1	4.2	
Kuwait	339.9	10.3	⇩	366.6	7.6	362.6	3.3	
Australia	495.2	30.3		522.2	5.2	475.8	5.5	⇩
Qatar	276.6	8.6	⇩	314.7	2.0	298.4	3.7	⇩

On average and across all countries, students in the low category achieved 27 score points below that of students in the medium category. Again, when focusing only on the countries where there is a statistically significant difference, the average difference between the students in the low and medium categories amounts to even 38 score points.

Table 4D.7 shows the same results for first-generation immigrant students. There are four countries where the mathematics achievement of students whose mathematics teachers put a high emphasis on mathematics homework is statistically significantly lower than of students whose mathematics teachers put a medium emphasis on mathematics homework. There are four countries where the mathematics achievement of students whose mathematics teachers put a high emphasis on mathematics homework is statistically significantly better than the achievement of students in the medium category.

The average achievement difference across all countries between students in the high category and students in the medium category is six score points. For the countries where there is a statistically significant difference, the average difference between students in the high and medium category is nine score points. Again, the achievement difference between students in the medium and low category is on average higher.

There are three countries where first-generation students in the low category achieved at a lower level than first-generation immigrant students in the medium category. But there are 14 out of 53 countries where first-generation students in the low category achieved statistically significantly below that of first-generation students with mathematics teachers with a medium emphasis on mathematics homework.

Across all countries, the students in the low category achieved on average 39 score points below that of the students in the medium category. Again, when focusing only on the countries where there is a statistically significant difference, the average difference between the students in the low category and in the medium category amounts to even 49 score points.

Given the much smaller sample sizes and the accordingly higher sampling errors for the achievement results of first-generation immigrant students compared to native students, it is surprising that we can determine so many countries where the achievement of first-generation immigrant students in the low category is statistically significantly lower than the achievement of students in the medium category.

These differences are also higher than the ones found for native students. This seems to indicate that infrequent and only little homework is related stronger to immigrant students' than native students' mathematics achievement. Considering that the overall mathematics achievement of first-generation immigrant students is lower than the achievement of native students, this seems to contradict the results found by Hattie (2008, p. 235) who stated "The effects [of homework] are greater for higher than for lower ability students…" but rather supports Trautwein et al. who stated: "This interaction effect indicates that low-achieving students gain more than high-achieving students from extensive homework assignments" (Trautwein et al., 2002, p. 45). Further research on the effect of homework on students with an immigrant background is needed and could help us understand the effect that I described.

Table 4D.7 Mathematics achievement of first-generation immigrant students by mathematics teachers' emphasis on homework

Country	High			Medium		Low		
	Math ach	SE		Math ach	SE	Math ach	SE	
Japan	702.5	22.3	⇧	559.2	28.7	521.1	26.1	
Thailand	425.6	55.6		346.7	39.7	NA		
Iran, Islamic Republic of	380.2	21.7		310.7	49.6	444.9	13.1	⇧
Canada (Quebec)	577.8	26.8	⇧	509.1	9.6	493.1	11.6	
Bulgaria	458.9	19.7	⇧	391.6	10.5	438.1	26.9	
Korea, Republic of	648.8	89.0		599.0	35.9	655.3	24.2	
Spain (Basque Country)	486.6	16.0		446.9	12.5	423.8	8.3	
Czech Republic	537.3	18.9		501.2	11.1	480.1	9.3	
Israel	472.2	8.3	⇧	436.5	15.8	371.7	41.8	
Georgia	386.6	15.7		351.6	16.4	335.8	32.6	
Scotland	507.4	42.6		476.8	19.9	429.1	12.0	⇩
United States (Massachusetts)	521.1	22.5		491.4	12.0	456.5	18.4	
Saudi Arabia	325.4	25.3		303.6	9.1	291.8	8.3	
United States	487.3	12.8		465.5	6.3	439.5	9.8	⇩
England	530.0	22.3		508.5	21.2	478.5	15.4	
Lithuania	464.8	20.6		446.6	9.8	407.7	21.0	
Hungary	481.7	31.0		465.7	18.5	433.5	77.5	
Armenia	515.8	23.4		501.3	13.6	492.5	19.1	
Romania	384.4	11.7		374.0	27.1	382.4	214.3	
Singapore	635.4	9.1		626.3	7.8	571.1	21.3	⇩
Morocco	322.6	20.4		315.4	10.4	332.5	21.3	
Slovenia	459.9	19.3		452.9	7.9	391.8	17.8	⇩
Tunisia	411.6	11.1		408.2	8.1	373.6	33.2	
Botswana	351.7	19.6		348.3	13.5	336.5	22.8	
Indonesia	349.3	10.2		347.0	8.3	353.6	9.6	
Sweden	464.4	13.5		464.4	13.2	454.1	8.3	
Turkey	401.2	24.4		402.6	27.0	409.1	41.4	
Cyprus	427.7	11.3		429.3	6.9	483.7	23.4	⇧
Mongolia	390.7	8.1		392.3	8.2	371.8	18.9	
Syria, Arab Republic of	361.5	8.0		365.2	11.3	366.2	11.3	
Malaysia	428.1	15.6		433.2	10.8	410.0	15.6	
Colombia	314.4	13.4		319.6	10.7	321.0	15.6	
Italy	449.2	6.7		455.0	11.1	NA		
Ghana	258.2	11.2		265.2	9.0	271.8	11.7	
Russian Federation	494.9	10.8		502.6	12.4	NA		
United States (Minnesota)	484.7	18.8		492.7	14.5	458.4	13.7	
Egypt	339.0	7.4		347.0	7.0	368.7	8.3	⇧
United Arab Emirates (Dubai)	472.4	16.4		482.6	5.6	504.4	13.0	
Hong Kong, SAR	564.5	12.5		576.2	9.2	492.1	21.6	⇩

(continued)

Table 4D.7 (continued)

Country	High Math ach	SE		Medium Math ach	SE	Low Math ach	SE	
Jordan	381.8	22.6		393.9	12.0	363.5	14.1	
Norway	441.0	8.3		453.1	6.5	426.3	11.8	⇩
Bosnia and Herzegovina	450.0	9.2		463.2	6.7	428.4	7.8	⇩
Canada (British Columbia)	526.3	19.2		540.7	8.8	524.4	17.0	
Ukraine	389.4	13.0		403.9	11.0	343.5	10.1	⇩
Serbia	446.2	16.7		461.1	13.2	462.4	14.3	
Lebanon	412.8	8.5		429.3	7.6	420.5	14.4	
Australia	498.2	29.1		516.7	10.0	478.7	12.4	⇩
Kuwait	310.0	9.4		332.8	19.4	333.9	7.5	
Malta	418.2	19.5		442.1	6.0	337.5	26.8	⇩
Oman	292.6	32.4		321.0	8.1	311.0	11.0	
Canada (Ontario)	514.0	11.8	⇩	546.2	8.4	512.2	11.9	⇩
Qatar	278.1	8.6	⇩	310.9	5.0	278.4	3.8	⇩
Chinese Taipei	491.1	18.6		524.2	16.5	474.1	12.6	⇩
El Salvador	270.2	17.4		307.1	11.1	309.9	18.8	
Palestinian National Authority	294.5	13.7	⇩	339.1	9.9	313.6	14.5	
Bahrain	326.8	18.5	⇩	389.0	6.6	360.0	10.4	⇩

> I conclude that in most countries there are no differences between native students and immigrant students when it comes to the emphasis of their teachers on homework. But I find again Singapore, British Columbia, and Quebec with differing results: the percentage of first-generation immigrant students in classes with a low emphasis on homework is statistically significantly smaller than the percentage of native students. In a good number of countries, I found a clearly positive relation between the emphasis on homework in mathematics and mathematics achievement. Interestingly, in quite a number of countries, the relationship seems to be even stronger for first-generation immigrant students.

Concentration of Immigrant Students

I will in the following focus on the distribution of immigrant students in classes. As discussed under "peer effects" in the literature review in Chap. 2, the concentration of the immigrant students across classes is often raised in public debates. As seen in Chap. 2, the results in research are quite ambiguous, but there seems to be some evidence that there is a negative effect – at least for immigrant students – when immigrant students are clustered together in classes.

For the following analysis, the number of first-generation immigrant students per mathematics class is calculated. Then the number of classes with a certain number of immigrant students is summarized.

Table 4D.8 shows the absolute number of mathematics classes sampled in TIMSS 2007 with the number of first-generation immigrant students. If there are ten or more first-generation immigrant students, then these classes are combined into the category "more than 9."

There is one restriction regarding this statistic. For larger classes, the chance of finding an immigrant student in the class is higher. At the system level, this means that in countries where larger class sizes are found, the number of immigrant students per class will also be higher. To avoid this effect, one could have calculated the percentage of immigrant students in the class, but this resulted in statistics that were difficult to interpret, and the statistics especially for smaller classes were affected quite strongly by even very few immigrant students in the class. Since the studies and articles cited in Chap. 2 are working with actual numbers, it was decided to accept the influence of the class size and use the actual numbers of immigrant students per class.

As can be seen in the table below, in most countries, there are very few classes with more than six or seven first-generation immigrant students. But in Bosnia and Herzegovina, Palestine, Ghana, Hong Kong, Oman, Qatar, Syria, Egypt, and Dubai, there is a good number of classes with ten or more first-generation immigrant students.

Interestingly, the number of classes with a different number of immigrant students varies quite substantially. In countries with a high percentage of immigrant students, there are classes with few immigrant students but also classes with a medium and high number of immigrant students. But also countries with similar percentages of immigrant students have quite an even distribution of immigrant students in the classes. For example, in Slovenia and Georgia, there are quite similar percentages of first-generation immigrant students as can be seen in Table 4A.1 (5 and 6 %, respectively). But whereas in Slovenia about half of the first-generation immigrant students are in classes with only one first-generation immigrant student, in Georgia about three quarters are in classes with two or more first-generation immigrant students.

Since we know now that there are differences within and between countries in terms of the distribution of immigrant students in classes, I will explore if this impacts the mathematics achievement results. For this I will calculate the achievement of all students and list the scores by the number of immigrant students in the mathematics class.[1]

---

[1] As discussed in chapter three, the within sampling units within schools in TIMSS are mathematics classes, and the class affiliation in this case reflects the mathematics classes. Since some of the educational systems are using course systems where the combination of students changes between their courses (e.g., the USA), having mathematics classes as sampling units means that the homeroom classes might be different. This is used in the analysis shown here.

Table 4D.8 Number of classes with specific number of first-generation immigrant students

Country	First generation immigrant students in the class										
	0	1	2	3	4	5	6	7	8	9	>9
Armenia	55	77	58	28	10	11	1	2	0	1	7
Australia	60	69	54	25	10	8	5	2	2	2	1
Bahrain	32	28	41	31	15	12	18	7	5	6	6
Bosnia and Herzegovina	12	25	21	16	21	16	17	11	5	3	34
Botswana	54	37	32	12	5	3	4	0	0	1	3
Bulgaria	89	76	35	26	9	8	1	1	2	0	0
Canada (British Columbia)	26	40	30	23	9	18	9	6	10	3	13
Canada (Ontario)	88	51	27	10	9	6	10	8	1	1	3
Canada (Quebec)	80	65	34	19	13	5	2	3	2	1	2
Chinese Taipei	33	50	31	25	10	2	1	1	0	0	0
Colombia	55	36	26	13	10	4	3	1	0	0	1
Cyprus	63	85	49	27	16	7	4	3	3	1	1
Czech Republic	140	45	17	2	6	0	1	1	0	0	0
Egypt	4	9	10	19	18	21	16	7	11	14	109
El Salvador	49	42	27	21	6	1	0	1	1	0	0
England	89	73	39	16	13	6	1	1	0	0	0
Georgia	70	52	38	12	8	1	3	0	0	0	0
Ghana	21	15	23	18	16	16	13	6	12	7	27
Hong Kong, SAR	2	3	6	9	12	19	14	8	9	9	29
Hungary	165	53	25	1	1	1	0	0	0	0	0
Indonesia	21	22	18	16	19	19	7	8	2	4	13
Iran, Islamic Republic of	178	23	6	1	0	0	0	0	0	0	0
Israel	25	42	22	20	9	11	4	3	2	2	6
Italy	151	78	41	12	3	2	0	0	0	0	0
Japan	134	26	5	1	0	0	0	0	0	0	3
Jordan	29	24	28	29	23	21	14	12	5	6	9
Korea, Republic of	134	14	0	2	0	0	0	0	0	0	0
Kuwait	9	24	21	22	22	22	10	11	6	2	9
Lebanon	26	28	34	32	21	20	8	11	4	6	15
Lithuania	140	80	31	6	1	0	0	0	0	0	0
Malaysia	51	37	27	13	9	3	5	3	5	2	8
Malta	74	75	44	20	8	4	4	3	0	0	0
Mongolia	15	21	22	22	24	13	9	9	7	5	5
Morocco	57	24	22	11	7	5	1	1	1	0	2
Norway	94	88	50	25	1	3	3	0	0	0	0
Oman	17	25	25	10	18	9	9	9	7	5	24
Palestinian National Authority	6	17	24	12	13	17	14	8	9	9	24
Qatar	10	13	27	20	31	34	29	36	19	18	51
Romania	177	60	19	7	2	1	0	0	0	0	0

(continued)

Table 4D.8 (continued)

Country	First generation immigrant students in the class										
	0	1	2	3	4	5	6	7	8	9	>9
Russian Federation	119	68	38	29	11	5	1	0	0	0	0
Saudi Arabia	34	27	27	24	24	20	20	7	5	4	12
Scotland	104	74	43	12	6	4	0	0	1	0	0
Serbia	85	73	33	22	6	4	1	2	1	0	0
Singapore	95	105	57	33	15	9	6	6	0	0	0
Slovenia	113	102	30	12	3	0	0	0	0	0	0
Spain (Basque Country)	85	29	15	8	4	3	1	3	0	2	1
Sweden	112	87	51	33	13	8	2	0	1	0	0
Syria, Arab Republic of	2	5	7	14	15	21	10	10	12	13	41
Thailand	128	16	6	0	0	0	0	0	0	0	0
Tunisia	68	58	19	16	8	0	0	0	0	0	0
Turkey	101	26	12	6	0	0	1	0	0	0	0
Ukraine	58	45	41	15	10	7	5	1	1	0	1
United Arab Emirates (Dubai)	2	9	9	11	7	9	10	8	9	3	76
United States	209	136	63	28	29	21	5	10	2	1	6
United States (Massachusetts)	31	23	14	10	5	4	7	1	2	0	0
United States (Minnesota)	46	33	12	5	2	1	1	1	0	0	1

Three groups are created to make the results easier to visualize and interpret. The first group includes students attending classes without immigrants, the second group includes students who are in classes with one or two immigrants, and the last group includes students attending classes with three or more immigrants. Then, the average mathematics achievement is calculated for each of the three student groups and compared to detect statistically significant differences. The reference group for the analysis is the group of students attending classes with one or two students, and the significant differences of the other groups are shown in Table 4D.9.[2]

There are several countries where the classes without immigrant students outperform classes with immigrant students. For example, in Chinese Taipei, the classes without immigrant students outperform the classes with at least one immigrant student by 20 score points, which on the other hand outperform classes with two or more immigrant students by another 15 score points – but not statistically significant.

Also in the United States, the classes with at least one immigrant student are outperformed by 25 score points, which on the other hand outperform classes with two or more immigrant students by another 28 score points. Both differences are

---

[2] Since classes with more students have a higher chance of including immigrant students than smaller classes if immigrants would be allocated to classes randomly and independent of the class size, the class sizes of the three groups of classes were calculated and compared to avoid this obvious source of bias since as shown before there are correlations between class size and achievement of the students.

Table 4D.9 Mathematics achievement by the number of immigrants in class

Country	No immigrants			1 or 2 immigrants			3 or more immigrants	
	Math Ach	SE		Math Ach	SE		Math Ach	SE
Egypt	387	55.1	⇨	433	15.6	⇩	390	3.5
United Arab Emirates (Dubai)	379	4.7	⇩	415	13.2	⇧	466	2.9
Singapore	566	7.2	⇩	593	6.0	⇧	629	8.2
Lebanon	446	8.5	⇩	471	8.0	⇩	441	5.4
Australia	481	6.7	⇩	499	5.7	⇨	504	9.7
Israel	455	12.6	⇨	468	7.5	⇨	460	6.8
Canada (British Columbia)	484	6.2	⇨	496	4.0	⇧	525	4.7
Palestinian National Authority	372	15.9	⇨	382	7.3	⇩	362	4.6
Saudi Arabia	336	4.8	⇨	346	4.5	⇩	323	3.5
Italy	475	4.6	⇨	483	3.7	⇨	480	8.8
England	508	8.9	⇨	516	7.0	⇨	515	13.3
Japan	567	2.7	⇨	575	12.3	⇨	563	15.9
Jordan	439	8.7	⇨	445	9.0	⇩	414	5.6
Morocco	383	4.3	⇨	389	5.8	⇩	370	6.9
Hungary	516	3.9	⇨	522	6.1	⇩	386	39.1
Bosnia and Herzegovina	442	12.2	⇨	447	5.9	⇧	460	3.1
Cyprus	462	3.3	⇨	466	2.3	⇨	465	3.2
Armenia	492	7.2	⇨	496	4.1	⇨	508	8.2
Russian Federation	512	5.0	⇨	515	5.8	⇨	498	9.1
Turkey	433	6.4	⇨	435	10.6	⇩	396	15.1
Serbia	488	4.4	⇨	490	4.7	⇩	464	7.6
Ghana	328	12.6	⇨	329	11.3	⇩	302	5.5
Bahrain	409	3.7	⇨	408	3.7	⇩	388	2.4
Iran, Islamic Republic of	403	4.1	⇨	402	11.5	⇨	423	0.0
El Salvador	342	4.3	⇨	341	4.7	⇨	336	7.6
Romania	464	5.3	⇨	463	6.9	⇨	416	26.2
Canada (Quebec)	527	4.3	⇨	525	5.6	⇨	528	12.0
Sweden	496	3.4	⇨	493	2.5	⇩	477	4.8
Norway	469	3.3	⇨	466	2.5	⇧	478	4.7
Oman	394	8.6	⇨	389	5.9	⇩	363	4.7
Canada (Ontario)	513	7.0	⇨	509	4.6	⇧	532	5.8
Czech Republic	505	3.1	⇨	500	5.4	⇨	489	6.4
Kuwait	367	13.4	⇨	360	4.2	⇩	349	3.2
Georgia	416	9.9	⇨	407	8.3	⇨	402	10.4
Slovenia	506	3.0	⇧	498	2.7	⇨	482	12.0
Tunisia	425	4.0	⇨	416	3.3	⇨	423	5.9
Ukraine	477	5.8	⇨	465	5.7	⇩	434	9.7
Lithuania	511	2.7	⇧	499	4.5	⇨	486	20.8

(continued)

Table 4D.9 (continued)

	No immigrants			1 or 2 immigrants			3 or more immigrants	
Country	Math Ach	SE		Math Ach	SE		Math Ach	SE
Scotland	496	5.6	⇨	482	6.0	⇨	473	14.5
Spain (Basque Country)	507	3.7	⇨	493	7.2	⇨	478	10.0
Botswana	371	3.4	⇧	354	3.0	⇧	373	8.7
Colombia	395	5.4	⇨	379	6.8	⇨	357	13.9
Indonesia	435	13.1	⇨	417	7.7	⇩	380	5.2
Chinese Taipei	619	6.8	⇧	596	6.3	⇨	581	7.1
United States	530	4.4	⇧	505	3.8	⇩	477	6.1
Malta	517	1.2	⇧	489	1.3	⇩	427	2.1
Hong Kong, SAR	622	0.8	⇨	592	23.6	⇨	569	6.0
Mongolia	478	10.2	⇧	445	7.4	⇩	423	4.4
Bulgaria	496	6.2	⇧	458	8.3	⇩	422	9.3
Malaysia	508	6.9	⇧	469	8.1	⇩	441	7.6
Syria, Arab Republic of	473	74.5	⇨	423	10.3	⇩	393	3.8
Qatar	374	4.8	⇧	320	2.1	⇩	304	0.8

statistically significant in the United States. The most extreme case is Hungary where the classes with more than two immigrants students are outperformed by classes with only one or two immigrant students by 136 points, and by 130 score points by classes without immigrant students; however, this result is based on three classes and should not be overinterpreted.

But there are also countries where the achievement does not differ significantly, as, for example, in Cyprus, England, or Italy. On the other hand, there are participants where classes with immigrants outperform classes without immigrants. In British Columbia, Canada, classes with at least one immigrant student outperform classes without immigrants by 12 score points and were outperformed by classes with two or more immigrant students by another 29 score points.

In Singapore classes with at least one immigrant student outperform classes without immigrants by 27 score points and were outperformed by classes with two or more immigrant students by another 36 score points. In Norway, classes without immigrants and classes with one immigrant score at about the same level (496 and 466 score points, respectively), but the classes with two or more immigrants achieved 478 score points which is significantly higher than the 466-point achievement of classes with one immigrant student.

Overall I find 14 countries with a statistically significantly lower mathematics achievement in classes with two or more first-generation immigrant students compared to classes with only one immigrant student. On the other hand, I find seven countries with a statistically significantly higher mathematics achievement in classes with two or more first-generation immigrant students compared to classes with only one immigrant student. And I find ten countries with a statistically significantly lower mathematics achievement in classes with one first-generation immigrant students compared to classes with no immigrant student. But then there are four countries with a statistically significantly higher mathematics achievement in classes with one first-generation immigrant student compared to classes with no immigrant student. In this group are Singapore and Dubai, two countries where the mathematics achievement of immigrant students is higher than the achievement of native students, as well as Lebanon and Australia, two countries where the mathematics achievement of first-generation immigrant students is lower than the achievement of native students (see Chap. 4A). It would be interesting to learn by what this is caused.

This analysis seems to support the findings of previous research that there is a tendency that students in classes with a significant fraction of immigrant students achieve lower, but there are exceptions. Since the analysis does not distinguish between the achievement of immigrant and native students in the class, the differences are highly influenced by the achievement differences between immigrant and native students. The following analysis shall explore this in more depth.

To further examine the effect of the number of immigrant students in class on the achievement, the mathematics achievement was calculated separately for immigrants and nonimmigrants by the three groups of classes: classes without immigrants, classes with one or two immigrants, and classes with more than two immigrant students.

First, the effect on immigrant students shall be evaluated. As discussed in Chap. 2, so far research has shown that a high percentage of immigrant students in a class has a negative peer effect on the achievement of immigrant students – mostly regarding the language achievement.

Table 4D.10 shows the mathematics achievement of students in classes with one or two immigrant students in comparison to classes with two or more immigrant students. In Botswana, British Columbia, Dubai, and Singapore, students in classes with more than two immigrant students significantly outperform immigrant students in classes with only one or two immigrant students.

In 11 countries, the opposite is the case, with Hungary being the most extreme example where immigrant students in classes with only one or two immigrants outperform immigrant students in classes with more than two immigrants by 140 score points. One must keep in mind, however, that there are only three classes with more than two immigrants in the Hungarian sample.

In Malta and Japan, the mathematics achievement difference between immigrant students in classes with only one or two immigrants and immigrant students in classes with more than two immigrants also exceeds 70 score points, but in Japan, only one class has more than three immigrant students.

There seems to be a tendency that a higher percentage of immigrant students and lower achievement of immigrant students go together. Yet there are examples where the mathematics achievement of immigrant students is higher when they are taught together with other immigrant students in the class.

Now that I examined the achievement differences for immigrant students, I will explore if there are any differences to be observed for native students depending on the number of immigrant students in the class. Table 4D.11 shows the mathematics

Table 4D.10 Mathematics achievement of immigrant students by the number of immigrants in class

Country	Classes with 1 or 2 immigrants		Classes with more than 2 immigrants		Difference		
	Math Ach	SE	Math Ach	SE	Math Ach	SE	
Hungary	481	13.0	341	40.6	140	42.7	⇧
Japan	543	19.7	464	22.4	78	29.8	⇧
Malta	467	8.5	395	8.2	71	11.8	⇧
Romania	394	12.8	343	23.5	50	26.7	⇨
Palestinian National Authority	372	18.5	325	7.6	47	20.0	⇧
Egypt	394	19.7	351	4.9	42	20.3	⇧
Colombia	343	9.8	303	12.9	40	16.3	⇧
Turkey	414	19.5	378	30.2	35	36.0	⇨
Indonesia	382	13.6	348	6.1	33	14.9	⇧
Qatar	327	14.9	294	3.0	33	15.2	⇧
Ghana	293	14.2	261	6.0	32	15.4	⇧
Bulgaria	426	13.0	394	13.5	31	18.7	⇨
Lebanon	448	11.6	419	5.3	29	12.7	⇧
Oman	342	16.2	313	6.5	29	17.4	⇨
Hong Kong, SAR	580	31.9	552	8.4	27	33.0	⇨
Serbia	468	13.7	442	11.5	26	17.9	⇨
Saudi Arabia	319	10.7	295	7.0	24	12.8	⇨
Morocco	337	10.2	313	8.7	24	13.4	⇨
Malaysia	447	13.0	423	9.7	24	16.2	⇨
Syria, Arab Republic of	387	15.4	363	6.0	24	16.5	⇨
Ukraine	411	11.3	388	12.3	23	16.7	⇨
Sweden	469	7.0	447	8.2	22	10.8	⇧
United States	479	6.1	461	7.4	18	9.6	⇨
Israel	456	15.7	439	9.4	18	18.3	⇨
Jordan	398	12.2	382	10.3	17	15.9	⇨
Scotland	456	10.9	440	18.6	17	21.6	⇨
Lithuania	448	9.7	432	21.7	16	23.7	⇨
Spain (Basque Country)	455	14.2	443	11.4	12	18.2	⇨
El Salvador	303	10.6	291	17.0	12	20.0	⇨
Bahrain	385	11.8	375	4.7	10	12.6	⇨
Mongolia	397	12.5	388	6.1	9	13.9	⇨
Georgia	374	15.6	366	21.2	9	26.4	⇨
Slovenia	449	8.0	443	17.2	6	19.0	⇨
Canada (Quebec)	517	8.9	512	12.2	5	15.1	⇨
Russian Federation	501	9.8	498	13.8	3	16.9	⇨
Tunisia	407	6.5	404	11.4	3	13.2	⇨
Iran, Islamic Republic of	376	20.2	376	22.4	0	30.2	⇨
Kuwait	326	11.5	329	6.8	−3	13.4	⇨

(continued)

Table 4D.10 (continued)

Country	Classes with 1 or 2 immigrants		Classes with more than 2 immigrants		Difference		
	Math Ach	SE	Math Ach	SE	Math Ach	SE	
Chinese Taipei	490	11.8	500	14.5	−9	18.7	⇨
Italy	449	7.4	459	9.3	−10	11.9	⇨
Norway	441	5.4	452	7.7	−11	9.4	⇨
Czech Republic	483	10.3	494	9.9	−11	14.3	⇨
Australia	488	8.8	503	11.5	−15	14.5	⇨
England	487	10.2	502	18.1	−15	20.8	⇨
Cyprus	417	9.0	435	6.7	−18	11.3	⇨
Canada (Ontario)	517	10.6	540	6.9	−23	12.6	⇨
Bosnia and Herzegovina	426	14.3	455	4.8	−29	15.1	⇨
Singapore	602	7.9	639	9.1	−37	12.1	⇩
Armenia	483	10.1	523	19.4	−40	21.8	⇨
United Arab Emirates (Dubai)	442	17.9	489	3.5	−47	18.2	⇩
Canada (British Columbia)	493	10.1	541	8.5	−49	13.3	⇩
Botswana	312	11.4	380	15.0	−68	18.8	⇩

Table 4D.11 Mathematics achievement of nonimmigrants by a number of immigrants in class

Country	No immigrants in class		Difference	1 or 2 immigrants in class		Difference	More than 2 immigrants in class	
	Math Ach	SE		Math Ach	SE		Math Ach	SE
	Math Ach	SE		Math Ach	SE		Math Ach	SE
Egypt	373	56.4	−64⇨	437	18.2	14⇨	423	3.5
Lebanon	445	8.7	−29⇧	474	8.0	18⇩	456	5.3
Singapore	567	7.2	−25⇧	592	6.1	−34⇧	627	8.4
Australia	483	7.1	−19⇧	502	5.9	−4⇨	505	9.8
United Arab Emirates (Dubai)	382	8.4	−14⇨	396	10.8	−50⇧	446	3.9
Canada (British Columbia)	484	6.4	−12⇨	496	4.1	−25⇧	521	3.9
Japan	569	2.9	−12⇨	580	12.3	25⇩	555	3.6
Italy	476	4.7	−12⇨	487	3.8	2⇨	486	10.3
Jordan	438	8.8	−12⇨	450	9.2	26⇩	424	5.9
Saudi Arabia	337	4.9	−11⇨	348	5.4	16⇩	331	3.5
Morocco	383	4.5	−10⇨	393	5.3	16⇩	377	6.7
England	511	9.1	−9⇨	520	6.9	0⇨	519	12.5

(continued)

Table 4D.11 (continued)

Country	No immigrants in class		Difference	1 or 2 immigrants in class		Difference	More than 2 immigrants in class	
	Math Ach	SE		Math Ach	SE		Math Ach	SE
Hungary	517	3.9	−9⇒	526	6.1	115⇓	411	37.4
Palestinian National Authority	372	16.3	−9⇒	381	7.3	2⇒	379	4.8
Cyprus	464	3.5	−8⇑	472	2.3	−4⇒	476	3.6
Israel	466	12.6	−7⇒	473	7.4	−2⇒	475	6.5
Ghana	326	12.8	−7⇒	332	11.4	16⇒	317	5.8
Romania	464	5.3	−6⇒	470	6.7	36⇓	435	26.7
Bosnia and Herzegovina	443	12.6	−6⇒	449	5.9	−15⇑	464	3.5
Serbia	489	4.4	−4⇒	494	4.7	25⇓	468	7.6
Turkey	433	6.5	−4⇒	437	10.6	44⇓	393	14.9
Russian Federation	514	5.1	−3⇒	518	5.8	17⇓	501	9.2
Armenia	495	7.5	−3⇒	498	4.0	−6⇒	504	6.3
Iran, Islamic Republic of	404	4.3	−1⇒	405	11.5	−10⇒	415	13.1
Sweden	497	3.5	0⇒	497	2.6	9⇓	488	4.8
El Salvador	344	4.6	1⇒	343	5.0	0⇒	343	7.1
Bahrain	411	4.1	2⇒	410	3.8	17⇓	393	2.7
Canada (Quebec)	531	4.6	2⇒	529	5.3	−6⇒	535	12.3
Norway	472	3.4	2⇒	470	2.6	−15⇑	484	5.3
Georgia	418	10.4	4⇒	414	8.4	6⇒	408	10.6
Slovenia	508	3.2	5⇒	503	2.7	10⇒	493	11.9
Czech Republic	507	3.2	5⇒	502	5.4	13⇓	488	6.0
Canada (Ontario)	515	7.1	6⇒	509	4.6	−20⇑	529	6.1
Oman	395	7.9	6⇒	389	6.4	12⇒	377	5.0
Ukraine	477	5.9	6⇒	471	5.6	22⇓	448	9.2
Lithuania	512	2.8	6⇒	506	4.3	7⇒	499	20.7
Kuwait	371	13.8	7⇒	364	4.5	6⇒	358	3.4
Tunisia	425	4.1	7⇒	417	3.5	−9⇒	426	5.6
Spain (Basque Country)	508	3.8	10⇒	498	7.0	4⇒	494	9.6
Scotland	498	5.7	11⇒	487	5.9	1⇒	486	13.4
Botswana	372	3.6	13⇓	358	3.0	−14⇒	372	8.8
Colombia	395	5.5	13⇓	382	6.7	17⇒	366	14.1
Chinese Taipei	621	6.9	17⇓	604	6.3	10⇒	594	7.0
United States	531	4.4	22⇓	509	3.7	27⇓	482	6.1
Indonesia	442	14.6	22⇒	420	7.9	25⇓	395	5.3

Table 4D.11 (continued)

Country	No immigrants in class		Difference	1 or 2 immigrants in class		Difference	More than 2 immigrants in class	
	Math Ach	SE		Math Ach	SE		Math Ach	SE
Malta	519	2.1	27⇩	491	1.7	53⇩	438	2.7
Mongolia	479	10.9	30⇩	449	7.7	18⇩	431	4.5
Bulgaria	498	6.1	34⇩	465	8.5	27⇩	437	8.8
Hong Kong, SAR	629	9.1	35⇩	594	23.6	16⇨	578	5.8
Malaysia	508	6.8	37⇩	471	8.0	25⇩	446	7.8
Syria, Arab Republic of	469	74.6	49⇨	420	11.4	16⇨	404	3.9
Qatar	373	6.8	54⇩	319	3.3	11⇩	308	2.1

achievement of nonimmigrant students by classes with none, one or two, and more than two immigrant students. Mostly there seems to be a negative peer effect on native students if the number of immigrant students in the class increases. This is in line with most previous research with the difference that there seems to be a stronger relationship for native students than for immigrant students.

There are, however, interesting counterexamples regarding peer effects. In Singapore, Australia, Dubai, and British Columbia, native students who are in classes with more than two immigrants perform significantly better than their peers in classes with fewer or no immigrant students. Since in Singapore, Dubai, and British Columbia immigrant students outperform nonimmigrant students, there seems to be a peer effect in favor of nonimmigrant students who are taught together with the high-performing immigrant students. This effect of higher-performing students who have a positive effect on their peers in class is also known as reflected glory effect as described in Chap. 2.

These positive effects might invite the interpretation that peer effects are less related to student's immigrant backgrounds but rather to student's achievement level. In that sense, increasing immigrant students' achievement can be regarded as a measure to increase the achievement of their native peers.

All these statements must of course be treated with caution. The data is cross-sectional and there is no experimental design. There might be selection effects and other unknown hidden factors. While I cannot, therefore, examine real effects, the results might still give an indication of potential effects.

I conclude that there are differences between countries in terms of the distribution of immigrant students across classes. There are some countries with a very high number of immigrant students in the class, and I detected a tendency that students in classes with immigrant students perform less well in mathematics. But the picture is ambiguous. For the mathematics achievement of first-generation immigrant students, I found that in a couple of countries (11), the mathematics achievement of immigrant students is higher when there are only one or two immigrant students in the class. For native students, there also seems to be a negative association between their mathematics achievement and the number of immigrant students in their classes, and it is even stronger than for immigrant students. But the results for those countries where the mathematics achievement of immigrant students is higher than of native students suggest that the effect is more related to peers with lower performance than to peers who are immigrants. Of course I could not measure causal effects and I use the term "effect" in a statistical sense here.

Summary

There is some evidence that smaller classes have a positive impact on student achievement – which is mostly supported by the STAR project conducted in the 1980s in the United States. For most countries, I cannot find any differences in percentages of immigrant students and native students attending smaller classes. But there are some exceptions in both directions. In Bahrain, Bulgaria, Malta, Qatar, Romania, and Scotland, there is a statistically significantly higher proportion of first-generation immigrant students in smaller classes than native students. In the United States, there are statistically significantly more native students in small classes than first- and second-generation immigrant students. Also in Jordan and Norway, there are statistically significantly more native students in small classes than second-generation immigrant students.

In terms of the overall relationship between class size and mathematics achievement, I could not find a positive association between smaller classes and higher achievement. Counterintuitively the mathematics achievement of immigrant students in smaller classes tends to be lower than that of their peers in larger classes. Related factors like smaller classes may be found more often in rural areas where students tend to achieve lower, but the data is not large enough to allow for a more in-depth analysis of this phenomenon.

For the emphasis on homework, there is no difference between teachers of immigrant students and native students except for the cases of Singapore and two of the Canadian provinces where there are less first-generation immigrant students in classes with a lower emphasis on homework. The effect on the mathematics achievement is similar for native and immigrant students: Students in classes with a higher emphasis on homework tend to achieve higher in mathematics.

Taking into account the number of immigrant students per class, we can see that in most countries, there are very few classes with more than six or seven first-generation immigrant students. Only in Bosnia and Herzegovina, Palestine, Ghana, Hong Kong, Oman, Qatar, Syria, Egypt, and Dubai, there is a higher percentage of classes with ten or more first-generation immigrant students.

There is also a tendency that students in classes with more immigrant students tend to perform lower. Counterexamples are Singapore, Dubai, and British Columbia where immigrant students achieve relatively high compared to native students and there is a positive effect on mathematics achievement.

So far I found immigrant students to achieve lower in most countries that I examined. I found some interesting results regarding the students' background as well as the school and class characteristics for native and immigrant students. These will be summarized and discussed in the summary at the end of the publication. But I also found a few countries where immigrant students achieved higher in mathematics. In the following two chapters, I will look deeper into the cases of Singapore and Canada to find potential reasons for the higher performance.

Chapter 5A
Immigrant Students in Singapore

Abstract From the beginning of its independence in 1965, Singapore has had a quite diverse and multicultural population. Unlike other countries, immigrant students in Singapore show consistently high achievement scores and – with the only exception of the TIMSS 2003 cycle – achieve the same level as native students, or even outperformed their native peers. This chapter seeks to explore the difference in approach towards immigrant students that Singapore takes in comparison to other countries. The history of Singapore's education system that ranks high in all international large-scale system monitoring surveys is described and compared to other international education systems. The chapter shows statistics for the achievement of first- and second-generation immigrant students and sheds some light on background variables such as the language spoken at home and parental background. Singapore subjects students with an immigrant background to high restrictions when it comes to school access and the consequences are highlighted in detail. On the other hand, Singapore not only has various support programs for immigrant students, some of which will be presented here, it also has policies in place fostering positive attitudes towards immigrants in general.

Keywords Education in Singapore • Immigrants in Singapore • Learning support program • Enhanced learning support program

The History of the Educational System in Singapore

Singapore gained independence in 1959 and separated from Malaysia in 1965. In 1965 the literacy rate in Singapore was 60 % (Chong & Cheah, 1997). Singapore has no natural resources or agriculture or industry. As Yue has shown, less than 2 % of the population was working in the agricultural and fishery industry in 1980–1999; mining or other industries gaining from natural resources are not listed at all (Yue, 2001, p. 28).

The Singaporean policy makers understood that their main resource is the country's population, which emphasized the need for it to be an educated one. "Apart from people, we have no other natural resources, hinterland or agriculture. Our livelihood depends on enterprise and hard work. It depends on our wits too, and

© Springer International Publishing Switzerland 2016   187
D. Hastedt, *Mathematics Achievement of Immigrant Students*,
DOI 10.1007/978-3-319-29311-0_8

our ability to adapt quickly every time the environment changes… to compensate for Singapore's natural resource deficiencies, (the government) emphasized the human factor: policies were designed to affect the behavior of people and to maximize their individual potential and contribution to the country" (Chong & Cheah, 1997, p. 2; Goh, 2005). This resulted in big investments in education and said policies.

Immigration in Singapore

From the beginning of its independence, Singapore has had a quite diverse and multicultural population. In 1970 77.0 % of the population was Chinese, 14.8 % Malays, 7.0 % Indians, and 1.2 % others (Department of Statistics, Singapore, 2013). The Singaporean society was defined by a mixture of cultures; therefore, in order to avoid conflicts due to different cultural backgrounds, but rather see the challenge of multiculturalism as a chance, policies were implemented to avoid conflicts. Even today the government is taking measures to mix cultures rather than foster the creation of subgroups within the population. For example, when new areas are created for settlements or when existing settlements are enlarged, the Ethnic Integration Policy determines that a certain percentage of each ethnic group should live there. If a certain percentage of inhabitants from an ethnic group is living in one area, "… transactions that make the community more segregated will not be allowed" (Wong, 2013, p. 6). Moreover the schools usually have a mixture of the cultures as intakes. In school lessons not only the various cultures are respected, but efforts are undertaken to create multicultural groups and to learn and benefit from each other. An example for this is music groups that are formed in various schools that use traditional instruments from all three cultural groups and combine them in one orchestra.

All these measures have probably had a positive impact on the acceptance of new immigrants as well. But still some resistance of the Singaporean population on further immigration was observed which then impacted immediate actions from the policy's side to reduce this effect (Sidhu, Ho, & Yeoh 2011).

Immigrants are an important factor for the Singaporean economy. The number of births per female Singaporean declined from 4.66 in 1965 to 1.24 in 2006 (Chong & Cheah, 1997). This results not only in an aging population but also in a reduced workforce stemming from native Singaporeans. Singapore is actively recruiting people from other countries with special emphasis on well-educated immigrants. The Singaporean law regulates the immigration. "Skilled workers and professionals and entrepreneurs are encouraged to take up a permanent residence and citizenship may be grated after 2–10 years of residence. Unskilled foreign workers, on the other hand, are permitted to work only for a limited time period, after which they are expected to return home" (Brownlee & Mitchell, 1997). To summarize this in Zygmunt Baumann's terminology (Baumann, 2003), Singapore encourages the tourists to come to the country and even to apply for permanent residency but allows vagabonds to come into Singapore only to work for a limited time.

Singapore's Education System in International Comparison

The Singaporean education system ranks high in all international large-scale system monitoring surveys. This pattern is observed not only in single points in time, but it is a fact that is repeatedly shown in surveys. In IEA TIMSS 1995 Singapore ranked first in mathematics and science in grade eight and ranked first in mathematics and seventh in science in grade four (Beaton, Martin, et al., 1996) (Beaton, Mullis, et al., 1996) (Mullis, et al., 1997) (Martin, et al., 1997). In TIMSS 1999 Singapore ranked first in mathematics and second in science (Mullis, Martin, Gonzalez, et al., 2000) (Martin, Mullis, Gonzalez, et al., 2000). In TIMSS 2003 Singapore ranked first in mathematics and science in both grades four and eight (Martin, Mullis, Gonzalez, & Chrostowski 2004) (Mullis, Martin, Gonzalez, Gregory, & Chrostowski 2004). In TIMSS 2007 Singapore ranked second in mathematics in grade four and third in grade eight but in both cases not statistically significantly below the first- and second-ranked country. For grade eight this was caused by a slight drop in the achievement of 13 score points compared to TIMSS 2003. For grade four the decrease in the ranking was despite the fact that the Singaporean achievement increased slightly (Martin, Mullis, Gonzalez, et al., 2004) (Mullis et al., 2004). In TIMSS 2011 Singapore again had the highest mathematics achievement for grade four students and the second highest for grade eight students – not statistically significantly below first-ranked Korea. This improvement was caused by a slight increase of the grade four achievements and a remarkable and statistically significant increase of the grade eight performance of nearly one fifth of a standard deviation. In TIMSS 2011 in science, Singapore ranked first in grade eight – statistically better than all other countries – and second in grade four, not statistically significantly below first-ranked Korea. The grade eight student achievement in science was 23 score points higher than in TIMSS 2007 – an increase of nearly a quarter of a standard deviation (Martin et al., 2012; Mullis et al., 2012). In summary, one can conclude that the achievement of Singaporean students in mathematics and science always was and is in the top ranks and is still increasing.

Immigrant Students in Singapore

I will now summarize the results from the analysis of the TIMSS 2007 data for Singapore. As can be seen in Tables 4A.1 and 4A.3, there are 11 % of first-generation immigrant students and 19 % of second-generation immigrant students in Singapore. Table 4A.4 shows that not only native students with an average mathematics achievement score of 588 are among the highest performers – first-generation immigrants even outperformed them with an achievement of 622 score points. Second-generation immigrant students also performed statistically significantly higher than native students with an average of 597 score points. First-generation immigrant students are on average 10 months older in grade eight than native students (15.1 compared to 14.3 based on Table 4B.1), whereas second-generation immigrants are of the same age as native students. Considering the results from Cliffordson and Gustafsson (2010) regarding the effect of maturing compared to schooling in terms of student achievement, the higher achievement of first-generation immigrant students is put into perspective to a certain extent. These results cannot, however, fully explain the great magnitude of the achievement gap.

As for the age of first-generation immigrant students at the time of immigrating to Singapore, we see that 38 % of the students immigrated at the age of 10 or above, 24 % at an age between 5 and 10, and 38 % before the age of 5 (see Table 4B.2). The highest mathematics achievement of 631 score points is observed for those students who came to Singapore at the age of 10 or later, the second highest for students who came before the age of 5 with an average of 624 points, and the lowest for students who came between the age of 5 and 10 (see Table 4B.2). But due to great variance within these groups and consequently big standard errors, only the difference between students who immigrated after the age of 10 and those who immigrated between the age of 5 and 10 differs significantly, favoring those students who came to Singapore at a higher age. This is an interesting result since it is opposed to the theory that an earlier start in the educational system results in better achievement (see (Myers et al., 2009), which seems not to be true in Singapore.

When looking at the boy-girl differences, we observe that the participation rate is almost the same for boys and girls among native students as well as among the immigrant student populations. Among native students, girls are outperforming boys in mathematics achievement by 15 score points, among first-generation immigrants by 24 score points, and among second-generation immigrants by six score points, the latter difference being not statistically significant. This means that the highest-achieving group is first-generation immigrant girls and the lowest is native boys. The difference between these two groups is more than a quarter of a standard deviation, and when looking at the grade seven-grade eight differences in TIMSS 1995, it equals the learning gain of nearly one school year.

Focusing on the language that the students speak at home, I find that 56 % of native students, 51 % of first-generation immigrant students, and 67 % of second-generation immigrant students do not speak the language of the test at home. This is a very high percentage at first glance but less surprising after an in-depth examination. As we see in exhibit 4.1 (Olson et al., 2008, p. 67), Singapore tested all students in TIMSS 2007 only in English. As explained above, the population of Singapore consists mainly of Chinese but also of Malays, Indians, and a small group of other origins. In Singapore there are four official languages: English, Mandarin, Malay, and Tamil. Mandarin, Malay, and Tamil are regarded as the languages of heritage and culture (Silver, 2005), but "English was designated as the first language of the school and so it came to be referred as the "first language"." As a result, mother tongues were described as "second languages" (Silver, 2005, p. 55). This creates the curious effect that the language that children are learning first becomes their second language in later life. A big emphasis in Singapore is put on bilingualism – or in recent times even multilingualism. The literacy rate in 2010 was 96 %; 80 % of the Singaporean population is literate in English and 71 % of the population is literate in two or more languages (Department of Statistics, Singapore, 2013). In school, mathematics and science are taught in English; therefore, Singapore decided to test their students' mathematics achievement in TIMSS in English. The fact that English is a language that most of the students start learning only in school makes the high achievement of Singaporean students in an international assessment even more surprising and respectable. For good job opportunities and communication across ethnic groups, English is the only tool. "English was treated as a necessity with regard to inter-ethnic communication and economic development"(Silver, 2005, p. 53). Only recently, due to the increasingly open economic market in China has Mandarin also been regarded as playing a role in economics and the school policies changed accordingly. Singapore has a clearly streamed school system and only students with overall high achievement – and this means mainly mathematics, science, and English – were enrolled in the highest streams. In 2004 this was adapted and "[f]lexibility has been introduced in that students who fall into the lower stream for English and Math can take the regular mother tongue course rather than the simplified B Syllabus if their mother tongue scores merit this variation" (Silver, 2005, p. 60).

A very striking factor of Singapore's immigrant student is the parental background. Whereas only 33 % of native students and 26 % of second-generation immigrant students have parents with an ISCED level 5 education or higher, 64 % of first-generation immigrants have parents with an ISCED level 5 education or higher. On the other hand, only 10 % of first-generation immigrant students have parents with an educational level of ISCED 2 or lower compared to 19 % of second-generation immigrant students and 15 % of native students. As described above, Singapore is very selective when it comes to immigration into the country and there is a clear distinction between skilled and unskilled workers. This causes the majority of immigrants to have a very good educational background. Dr. Ng Eng Hen, Minister for Education and Second Minister for Defence, said in a speech at the 5th Teachers' Conference 2010 on September 6, 2010:"New immigrants, and those who become PRs and new citizens, have higher educational qualifications too. Last year, 2 in 3 new citizens came with post-secondary educational qualifications. The trend for PRs is similar—nearly 8 in 10 new PRs have post-secondary qualifications" (Hen, 2010). Previous research has shown, too, that the socioeconomic background of students explains a big portion of achievement differences between students (see, e.g., (Sirin, 2005).

Examining the Singaporean students' attitude, I can conclude that among all immigration groups, 85–86 % of the students like to go to school. But interestingly 75 % of native students answered that they like mathematics, as did 72 % of second-generation immigrants, but even 82 % of first-generation immigrants. The attitudes toward mathematics are statistically significantly higher for first-generation immigrants than for the other students. This result is coherent for boys and girls. The self-perception of the students' mathematics abilities also shows a similar pattern with 81 % of first-generation immigrant students reporting to do well in mathematics compared to 65 % of native students and 66 % of second-generation immigrant students. Obviously the students' self-perceptions match the achieved results across these three groups.

Access to Schools in Singapore

A factor that contributed to the positive achievement results of immigrant students is probably that students with an immigrant background must meet high requirements to access schools. Before they are allowed to enroll in a school in Singapore, they are required to pass an Admissions Exercise for International Students (AEIS) test. "AEIS consists of a centralized test on English and Mathematics that will assess the applicants' English literacy, numeracy and reasoning abilities. Applicants who pass the test will be offered a place in a suitable school, based on availability of school vacancies, their test performance and declared address in Singapore" (Ministry of Education, Singapore, 2013a). Since the requirements for the test are high, "International students are strongly encouraged to prepare themselves before taking the Admissions Exercise for International Students (AEIS) tests. They should be familiar with the English and Mathematics syllabi of the level preceding the one they are applying for" (Ministry of Education, Singapore, 2013a).

Support Programs in Singapore

A third factor for the success might be that students who have learning difficulties are strongly supported in the educational system. The learning support program (LSP) for English started in 12 primary schools in 1992. In 1996, the program was implemented in 93 primary schools and in 1998 the LSP was rolled out to all primary schools. "The LSP is a specialized early intervention program aimed at providing learning support to students who enter Primary 1 with weak English language and literacy skills. Students are identified for the LSP through a systematic screening process carried out at the beginning of Primary 1. The objective of the LSP is to equip students with basic literacy skills so that they could access learning in the regular classroom" (Ministry of Education, Singapore, 2013b).

"Each year, 12–14 % of the Primary 1 cohort is identified to require support in the LSP. Results have shown that the LSP has helped around 30 % of these children to read at their age level and pass their school-based English Language examinations by the end of Primary 1. Students who were not able to do so continued to receive support in Primary 2. At the end of Primary 2, another 10 % of students would have been able to read at their age level and pass their examinations" (Ministry of Education, Singapore, 2008).

The enhanced learning support program (ELSP) at Primary 1 level has been introduced to all primary schools from January 2007, while the program at Primary 2 level started out in 2008. 40 % of the students who underwent the ELSP met the discharge criteria at the end of Primary 1. The other 60% went on to the Primary 2 ELSP where 25 % met the discharge criteria, making it 65 % who were discharged at the end of Primary 2. The remaining 35 % were assessed for dyslexia or other forms of learning disability and were given appropriate and specific forms of support (Hong, 2009).

"The original LSP provided remedial help to students in what is being taught in the regular English classes. The enhanced LSP, on the other hand, teaches skills which can be used to learn in the different subject areas. It was designed for Primary 1 and 2 students, and focuses on five major components that international research has found to be critical in helping students in early primary grades learn to read and spell: Alphabetic knowledge, knowledge of sound-letter correspondences, fluency, vocabulary, and reading comprehension" (Ministry of Education, Singapore, 2008).

Other factors might relate to the general circumstances in Singapore. Since the population of Singapore is already multicultural and has a diverse background and since ethnic conflicts are a known problem in the Singaporean history, Singapore has policies in place to overcome them. Some of these policies are policies addressing the whole population, such as the policy for mixing ethnic groups in new settlements, or cultural festivals that help citizens become familiar with other ethnics' cultures. Others are addressing schools that are required to have a mix of cultures as intake and school bands supporting intercultural activities, e.g., consisting of students with different ethnical background playing classical instruments of their cultural origin. It can be assumed that new immigrants to Singapore are experiencing more openness than in other countries. Hence, integration in a multicultural population might be easier.

Summary

The Singaporean economy depends very much on human resources. Consequently, education has a high value, is in the focus of policy makers, and is getting very good financial support. As a result Singapore achieves high scores in international comparisons. The results for students with an immigrant background are also very positive. This has probably multiple reasons: Firstly, immigrants to Singapore have in general a very high level of education; consequently, the students entering the educational system have parents with a high socioeconomic status. Positive achievement resulting from this fact cannot, of course, be attributed to the educational system of Singapore. Secondly, immigrant students entering the educational system have to prepare in order to pass the admission test. Positive effects emerging from this can probably only be partially attributed to the Singaporean educational system's success and only in the sense that private initiatives to learn are strongly encouraged. The factor that might have a positive impact on immigrant students' achievement that can mostly be attributed to the Singaporean educational system is the invention of support programs for students with low achievement in mathematics and English. And last but not least, the attitude towards immigrants and the value of ethnic diversity are probably also fostering the integration of immigrants in society and in schools.

In summary I conclude that the Singaporean educational system is probably facing fewer challenges related to immigrant students than other countries although the number of immigrant students is relatively high. However, this may also be due to initiatives assisting students with an immigrant background.

Singapore is an example where the challenges of immigration for the educational system seem to be reduced and met quite successfully, thus somewhat reducing the problems that students with an immigrant background are facing. In future research it would be interesting to determine more clearly what the impact of the different implemented policies is. For this purpose more longitudinal research projects and case studies targeting immigrants would be beneficial.

I conclude that the selective immigration policies of Singapore might have impacted the positive achievement results of their immigrant students, and this probably cannot help policy makers in other countries learn how to improve the achievement of their immigrant populations. Singapore's experience on how to deal with a diverse and multicultural society probably cannot be easily mirrored in other countries, but maybe some of the specific policies could be used to inspire other countries' policies. The support programs for immigrant students might also have contributed to the positive achievement outcome and could be drawn upon to inform other countries' policies. We cannot disentangle which of the different aspects had impact on the educational outcome of the immigrant students; maybe additional data, for example, a comparison of the achievement of students enrolled and those not enrolled in the support programs, might shed some light on the different effects.

Chapter 5B
Immigrant Students in Canada

Abstract Following the previous chapter on Singapore, this chapter presents Canada as another case of successful immigrant students in a multicultural society that may serve as a positive example for other countries. Today, about 250,000 immigrants migrate to Canada each year, and about 20 % of the Canadian population consists of immigrants. Similar to Singapore, Canada gives preference to immigrants with a higher education – especially in more recent times, basing the immigration admission program on a credit point system. This section describes immigrant regulation as well as legislation in Canada and the distribution of immigrants across the Canadian provinces in relation to immigrant student achievement, taking into account the differing language situation. Focusing on the differences found between the provinces, the chapter concludes with a close look at factors in quantitative research, including age at immigration, the language spoken at home, gender differences, the school environment, student's self-esteem, and initiatives targeted at immigrant students.

Keywords Canada • Immigrant population in Canada • Immigrant statistics for Canada • Decentralized support programs

The Educational System of Canada

Canada is a federal system with respect to immigration as well as education policies. Federal law is applied as well as legislation from the ten provinces and three territories in the North. According to Section 93 of the Canadian constitution, education falls under the provincial jurisdiction and consequently education policies vary between provinces. Beyond this there is also a federal influence by allocating financial resources from the federal government. On the website of the Official Languages and Bilingualism Institute (OLBI) at the University of Ottawa it reads:

> "Moreover, problems can arise between the provinces and central authorities when the latter use their spending power to intervene directly or indirectly in areas of provincial jurisdiction. For example, the federal government can use its financial clout to influence certain social services like pension funds, various areas of education (primary and secondary teaching, vocational education and universities), health and municipalities" (University of Ottawa, n.d.).

© Springer International Publishing Switzerland 2016
D. Hastedt, *Mathematics Achievement of Immigrant Students*,
DOI 10.1007/978-3-319-29311-0_9

Immigration in Canada

Immigration in the Canadian system is a shared responsibility between the provincial jurisdiction and the federal jurisdiction. Section 95 of the Canadian constitution states:

"In each Province the Legislature may make Laws in relation to Agriculture in the Province, and to Immigration into the Province; and it is hereby declared that the Parliament of Canada may from Time to Time make Laws in relation to Agriculture in all or any of the Provinces, and to Immigration into all or any of the Provinces; and any Law of the Legislature of a Province relative to Agriculture or to Immigration shall have effect in and for the Province as long and as far only as it is not repugnant to any Act of the Parliament of Canada."

Although Canada has been an immigration country for centuries, the Canadian immigration policy changed significantly in the past. Going back about 150 years, I find:

"According to the 1870–71 census, Canada's total population was 3.6 million. In addition to native peoples (about 136 000 in 1851) the 2 largest groups were the French (1 million) and the British (2.1 million). Excepting the Germans (203 000), other groups (Dutch, American blacks, Swiss, Italians, Spanish, Portuguese) were much smaller. During the next century, about 9.3 million people immigrated to Canada and, although many went on to the US or eventually returned to their native lands, by 1996 Canada's population had surpassed 29 million" (Dirks, 2006).

Today about 250,000 immigrants migrate to Canada each year, and about 20 % of the Canadian population consists of immigrants:

"Canada has welcomed, as you all know, an annual average of more than 250,000 immigrants since 2006, since our government came into office. This is the highest sustained level of immigration in Canadian history. But to ensure that immigration will fuel and drive our future prosperity, we need to select immigrants who are ready, willing and able to integrate into Canada's labour market and to fill roles where we have existing skills shortages. We have to make sure that the skilled immigrants we choose are the ones Canada needs and are the most likely to succeed when they arrive" (Alexander, 2013).

Similar to Singapore, Canada gives preference to immigrants with a higher education – especially in more recent times. The immigration admission program is based on a credit point system which determines preferences for immigration. Up to 100 points can be obtained by potential applicants and 69 credit points are required to apply for admission. Between 1985 and 2004, up to 15 points could be obtained for a vocational training, 12 for education, and 15 for language skills in English or French. In 2004 the credit system changed. From then on vocational education was not rewarded anymore, but the education was counted for up to 25 credit points. As Simmons described: "After these most recent changes, a person with a post-secondary diploma or degree involving two or more years of study automatically receives 20 points or roughly 30 % of the 69 points required for admission. Anyone with less than a high school studies completed would get 0 points on the education criteria" (Simmons, 2010, p. 95). The language expertise was increased to 25 credit points. This means that about half of the credit points could be obtained by a high education and the mastery of English or French. Or in other words, without the credit points in education and language competencies, a potential applicant will not be able to achieve the 69 credit points needed for admission (see Simmons 2010).

Another aspect that regulates the immigration flow is that since March 1992 immigrants are charged a fee. For a couple with two children, the total amount of the fees sums up to 3,200 Canadian dollars. For less resourced people this poses an obstacle to the immigration application. Moreover this policy favors immigrants with a higher socioeconomic status.

However, not only economic but admittedly also humanitarian aspects are driving the immigration policies of Canada. Chris Alexander, Canadian Minister of Citizenship and Immigration, said in a presentation on Canada's immigration policies in September 2013:

"We see the tide of humanity that has sought to escape conflict in Syria and other parts of the world, some of whom – the most vulnerable among whom – will of course be welcome in Canada" (Alexander, 2013).

And indeed, Canada is the country accepting the third most refugees in the world. The immigration fee, however, also applies to refugees. While some of the refugees are applying for immigration under refugee aid programs, others apply and are accepted by applying for immigration based on the credit point system that prefers skilled workers.

As in Singapore, in Canada the public opinion regarding immigrants is rather positive. "In terms of public opinion, Canadians have a more positive view of immigrants and immigration than do Americans and Western Europeans. They are less likely to view immigrants as "stealing" jobs or committing crimes, and the majority of Canadians view immigration as an opportunity, not a problem. Furthermore, only 17 % of Canadians think there are "too many" immigrants in their country, compared to 37 % of Americans and 59 % of the British" (Statistics Canada, n.d.).

This positive public opinion might also influence the living conditions for immigrants and might make it easier for them to integrate into the society. Students, for example, can integrate into the schools system more easily. It might also foster the willingness to accommodate the needs of students with an immigrant background more openly.

But as stated above the immigration policies are partly subject to provincial legislation which includes the immigrant policies, and in consequence the latter also differ between the provinces and territories. Table 5B.1 displays the number of immigrants for Canada in total as well as the individual Canadian provinces and territories and the country of origin of the immigrants based on the 2006 census data (Statistics Canada, 2007). Table 5B.2 displays the distribution of immigrants' countries and regions of origin for Canada and for each province and territory.

Table 5B.1 shows that in 2006 more than six million immigrants are living in Canada. Nearly 3.4 million of them are living in Ontario which is more than half of the total immigrant population in Canada. The second largest group of immigrants – with about 1.1 million – can be found in British Columbia, followed by Quebec that accommodates 850,000 immigrants. These three provinces cover nearly 90 % of the Canadian immigrant population.

Table 5B.1 Immigrant population by place of birth, by province, and by territory (2006 Census)

| | 2006 | | | | | | | | | | | | | |
Place of birth	Canada (number)	N.B.	Que.	Ont.	N.L.	P.E.I.	N.S.	Man.	Sask.	Alta.	B.C.	Y.T.	N.W.T.	Nvt.
Total — Place of birth	6,186,950	26,395	851,555	3,398,725	8,380	4,780	45,195	151,230	48,160	527,030	1,119,215	3,010	2,815	450
United States	250,535	8,655	26,575	106,405	1,400	1,255	7,960	7,090	5,425	28,325	56,560	600	235	40
Central and South America	381,165	845	78,010	216,640	395	150	1,020	17,765	2,545	31,160	32,455	60	95	20
Caribbean and Bermuda	317,765	470	80,835	211,380	145	90	980	4,085	725	10,365	8,575	15	60	30
Europe	2,278,345	10,835	306,515	1,307,885	4,040	2,465	22,565	62,545	21,615	187,675	349,410	1,675	915	195
United Kingdom	579,625	5,210	16,035	321,650	2,335	1,165	11,665	15,225	7,690	60,215	137,460	555	345	90
Other Northern and Western Europe	489,540	3,790	92,555	209,610	945	960	6,640	15,845	5,670	53,020	99,225	860	340	70
Eastern Europe	511,095	805	72,765	304,495	420	145	2,110	18,875	5,255	46,610	59,320	150	115	25
Southern Europe	698,080	1,030	125,165	472,130	345	190	2,150	12,605	3,000	27,830	53,400	110	115	15
Africa	374,565	1,225	123,990	164,795	560	165	2,125	7,660	3,540	35,525	34,575	75	285	35
Asia and the Middle East	2,525,160	4,095	233,000	1,376,595	1,780	485	9,910	51,490	13,860	225,410	606,730	515	1,165	115
West Central Asia and the Middle East	370,515	950	81,035	213,980	270	160	3,950	3,965	1,715	24,775	39,605	10	90	10
Eastern Asia	874,370	1,440	52,655	417,985	545	225	2,735	7,635	5,055	72,330	313,415	135	180	25
Southeast Asia	560,995	735	56,420	270,710	245	35	1,115	31,290	4,765	73,675	120,865	275	820	50
Southern Asia	719,275	965	42,890	473,915	725	65	2,115	8,600	2,335	54,630	132,845	90	70	30
Oceania and other countries	59,410	265	2,620	15,025	65	165	630	580	435	8,570	30,910	70	60	15

Source: 2006 Census of Population. Statistics Canada, 2007

As can be seen in Table 5B.2 in the different provinces and territories, there is a different distribution of source countries of immigrants. In British Columbia more than 50 % of the immigrants come from Asia – this is the highest percentage of Asian origin among the Canadian provinces and territories. Within the Asian group the majority of immigrants come from Eastern Asia (28 %). The second largest group of immigrants in British Columbia comes from Europe (31.2 %) with the majority originating from the United Kingdom (12.3 %).

The situation in Ontario is quite similar although there are fewer immigrants from Asia (40.5 %) and more immigrants from Europe (38.5 %), from Central and South America (6.4 %), and from the Caribbean (6.2 %). Compared to British Columbia the majority of the European immigrants' origin is Southern Europe and not the United Kingdom.

Somewhat different is the situation in Quebec. Whereas British Columbia and Ontario are – as all other provinces and territories – Anglophone, Quebec is mostly Francophone which reflects in that Quebec's immigrant population is mostly from Francophone areas. Notably only half as many immigrants as in British Columbia originate from Asia (27.4 %). The majority of Asian immigrants come from West-Central Asia and the Middle East (9.5 %), which is the highest number among all Canadian provinces and territories. Quebec also has the highest percentage of immigrants from Africa (14.6 %) and the Caribbean (9.5 %), among all Canadian provinces and territories, and the second highest percentage of immigrants from Central and South America (9.2 %). In terms of countries of origin, the immigrant population in Quebec seems to be more diverse than in Ontario and British Columbia and more affected by a common language background.

Another factor affecting the situation for immigrants in Quebec is that due to the language situation within Canada – with a minority of citizens speaking French – the Francophone citizens of Quebec are regarded as more isolated and linked closer together which might make it more difficult for foreigners to integrate into the society. Simmons (2010, p. 54) writes: "French speaking Quebecers in small towns are less familiar with immigrants in general, particular those of non-European origin. They live among ethnic kin with whom they are closely linked by intermarriage, have common religious practices, and a long history of living together." This might also be a reason for the fact that there are significantly less immigrants in smaller communities in Quebec (see Table 4.3.2). Whereas there are 14 % of immigrants in communities with more than 500,000 inhabitants and 6 % in communities with 3001–500,000 inhabitants, there are only 2 % in communities with 3000 or less inhabitants in Quebec. In Ontario and British Columbia, the percentage of immigrants is higher in general, but also in communities with less than 3,000 inhabitants, we find 5 % of immigrants in Ontario and even 9 % of immigrants in British Columbia.

Table 5B.2 Immigrant population by place of birth, by province, and by territory (2006 Census)

	2006													
	Canada	N.B.	Que.	Ont.	N.L.	P.E.I.	N.S.	Man.	Sask.	Alta.	B.C.	Y.T.	N.W.T.	Nvt.
	Percentage													
United States	4.05%	32.79%	3.12%	3.13%	16.71%	26.26%	17.61%	4.69%	11.26%	5.37%	5.05%	19.93%	8.35%	8.89%
Central and South America	6.16%	3.20%	9.16%	6.37%	4.71%	3.14%	2.26%	11.75%	5.28%	5.91%	2.90%	1.99%	3.37%	4.44%
Caribbean and Bermuda	5.14%	1.78%	9.49%	6.22%	1.73%	1.88%	2.17%	2.70%	1.51%	1.97%	0.77%	0.50%	2.13%	6.67%
Europe	36.83%	41.05%	35.99%	38.48%	48.21%	51.57%	49.93%	41.36%	44.88%	35.61%	31.22%	55.65%	32.50%	43.33%
United Kingdom	9.37%	19.74%	1.88%	9.46%	27.86%	24.37%	25.81%	10.07%	15.97%	11.43%	12.28%	18.44%	12.26%	20.00%
Other Northern and Western Europe	7.91%	14.36%	10.87%	6.17%	11.28%	20.08%	14.69%	10.48%	11.77%	10.06%	8.87%	28.57%	12.08%	15.56%
Eastern Europe	8.26%	3.05%	8.54%	8.96%	5.01%	3.03%	4.67%	12.48%	10.91%	8.84%	5.30%	4.98%	4.09%	5.56%
Southern Europe	11.28%	3.90%	14.70%	13.89%	4.12%	3.97%	4.76%	8.33%	6.23%	5.28%	4.77%	3.65%	4.09%	3.33%
Africa	6.05%	4.64%	14.56%	4.85%	6.68%	3.45%	4.70%	5.07%	7.35%	6.74%	3.09%	2.49%	10.12%	7.78%
Asia and the Middle East	40.81%	15.51%	27.36%	40.50%	21.24%	10.15%	21.93%	34.05%	28.78%	42.77%	54.21%	17.11%	41.39%	25.56%
West Central Asia and the Middle East	5.99%	3.60%	9.52%	6.30%	3.22%	3.35%	8.74%	2.62%	3.56%	4.70%	3.54%	0.33%	3.20%	2.22%
Eastern Asia	14.13%	5.46%	6.18%	12.30%	6.50%	4.71%	6.05%	5.05%	10.50%	13.72%	28.00%	4.49%	6.39%	5.56%
Southeast Asia	9.07%	2.78%	6.63%	7.97%	2.92%	0.73%	2.47%	20.69%	9.89%	13.98%	10.80%	9.14%	29.13%	11.11%
Southern Asia	11.63%	3.66%	5.04%	13.94%	8.65%	1.36%	4.68%	5.69%	4.85%	10.37%	11.87%	2.99%	2.49%	6.67%
Oceania and other countries	0.96%	1.00%	0.31%	0.44%	0.78%	3.45%	1.39%	0.38%	0.90%	1.63%	2.76%	2.33%	2.13%	3.33%

Source: 2006 Census of Population. Statistics Canada, 2007. Percentage calculated by the author

Immigrant Students in Canada

Table 4A.3 has revealed that in British Columbia the first-generation immigrant students' mathematics achievement in TIMSS 2007 is 33 score points higher than the mathematics achievement of native students – exactly the difference between grades eight and nine in TIMSS 1995 in Canada. Second-generation immigrant students in British Columbia performed 18 score points above native students in mathematics in TIMSS 2007. In Ontario first-generation immigrant students had 11 score points ahead of native students in TIMSS 2003 and 22 score points ahead in TIMSS 2007. Second-generation immigrant students in Ontario were at about the same level in mathematics as native students in TIMSS 2003 and TIMSS 2007 with three, respectively, nine score points ahead. In Quebec first-generation immigrant students had 30 score points behind native students in mathematics in TIMSS 2003, but the difference was reduced to only 17 score points in TIMSS 2007. Second-generation immigrant students in Quebec were only slightly behind with seven score points in TIMSS 2003 and one score point in TIMSS 2007.

In science the results for immigrant students were not that positive. In British Columbia first- and second-generation immigrant students achieved one score point better than native students in TIMSS 2007 – which was certainly within the measurement error. In Ontario first-generation immigrant students performed 15 score points below native students in science in TIMSS 2003 but one score point ahead in TIMSS 2007. Second-generation immigrant students in Ontario were eight score points behind their native peers in TIMSS 2003 and three score points ahead in TIMSS 2007. In Quebec first-generation immigrant students were 55 score points behind native students in science in TIMSS 2003 and still 21 score points behind in TIMSS 2007. Second-generation immigrant students in Quebec were 22 score points behind in TIMSS 2003 and still 13 score points behind native students in TIMSS 2007.

Statistics Canada conducted research on the language abilities of immigrant students with the following result: "The language skills of children of immigrant parents just entering the school system were weaker than those of Canadian-born parents, but the longer the children lived in Canada, the smaller the gap in performance became, until it disappeared. In fact, in later years, the academic performance of many of these students surpassed that of their Canadian-born counterparts" (Statistics Canada, n.d.).

The research from Worswick (2001) supported this observation. He concluded: "Overall, children of immigrants generally do on average at least as well as the children of the Canadian-born along each dimension of school performance. The children of immigrant parents whose first language is either English or French have especially high outcomes. The children of other immigrant parents have lower performance in reading than do other children; however, their performance in other areas is comparable to that of the children of Canadian-born parents. Evidence is also found that, with more years in the Canadian education system, the performance of these children in reading converges to that of the children of Canadian-born parents. In general, the results indicate that the children of immigrants have predicted performance in virtually all areas that is at least as good as the performance of the children of the Canadian-born by age 13. In a number of cases, this standard is met at much earlier ages" (p. 13).

In summary, I conclude that the situation is more positive for the immigrant population in mathematics than in science but that in both mathematics and science the trends are positive. Other research supported the positive results for students with an immigrant background in Canada.

Factors Found in the Quantitative Research

After investigating the general outcome, I am going to take a more in-depth look into the results from chapter 4 for Canada. My special focus will be on the differences between the provinces.

Examining the age immigrants at immigration, we observe that in British Columbia the immigrant students tend to immigrate at a later age than it is the case in Ontario. The immigrant students in Quebec were even younger when immigrating to Canada. In Quebec we see that the immigrant students that migrated at a younger age performed better than the students who migrated at a later age. In Ontario the result is exactly the opposite and the immigrant students who migrated later also score higher. In British Columbia the picture is ambiguous with no clear pattern.

When examining the differences between girls and boys, we find that only in Ontario there is a slightly smaller percentage of first-generation immigrant girls enrolled in schools (46 %). Also the mathematics achievement is very similar for boys and girls in the three Canadian provinces for all immigration groups and for native students. Only in Ontario first-generation immigrant students achieved significantly better than their female peers (543 versus 523 score points).

As already indicated above in terms of the language spoken at home, Chap. 4B revealed that in all three provinces that participated in TIMSS 2007, there are a very large number of first-generation immigrant students that do not speak the language of the test at home. In British Columbia 42 % of first-generation immigrant students do not speak English at home. In Ontario where most of the tests were conducted in English[1] (and some in French), 37 % of first-generation immigrant students do not speak the language of test at home; in Quebec where most of the tests were conducted in French (and some in English), 35 % of first-generation immigrant students

---

[1] It should be noted here that although the language of the test was primarily French in Quebec and English in Ontario and British Columbia, children in Quebec who were taught in English were also tested in English, and children in Ontario and British Columbia who were taught in French were tested in French. These were only small percentages in each of the provinces; however, we can conclude that the English minorities in Quebec and the French minorities in Ontario and British Columbia do not affect the results presented here.

do not speak the language of the test at home. These are among the highest percentages across all participants. Countries with a higher percentage are mostly countries that have also a high percentage of nonimmigrant students that do not speak the language of the test at home. In the Canadian provinces 6 % and less native students do not speak the language of the test at home. In Singapore, for example, 67 % of first-generation immigrant students do not speak the language of the test at home, but this is also true for 51 % of the native students. As described in Chap. 5A, in Singapore this is mostly driven by the fact that the population consists of three major ethnics groups with different native languages – Chinese, Malays, and Indians. In Ontario and in British Columbia, there are language support programs at a decentralized level. Schools are offering English Literacy Development (ELD) programs or English as a second language (ESL) programs. The programs can be integrated classroom programs, intensive or partial support programs, tutorial support, or other forms of support programs (Settlement.org, 2012; Province of British Columbia, 2013). However, the provinces are giving clear guidelines for these courses. In Ontario the Ministry of Education published: "The document provides practical strategies and models for integrating language and content instruction for ESL/ELD students in those classrooms" (Ministry of Education, Ontario, 2001, p. 6) and promotes a holistic approach that includes the parents, the teachers, as well as the whole school environment. It suggests:

Promoting an Inclusive and Supportive School Environment

All school staff members should work towards creating a welcoming and supportive atmosphere for ESL/ELD students. School administrators can help to create such an atmosphere by implementing some of the following suggestions:

- Post visual images that represent all students in the school.
- Provide signs, notices, and announcements in the languages of the school community.
- Honor the various cultural and faith celebrations within the school.
- Encourage and recruit bilingual volunteers.
- Have staff who provide ESL/ELD support collaborate in program planning.
- Promote professional development opportunities for ESL/ELD staff and classroom teachers.
- Take ESL/ELD considerations into account when creating timetables.
- Include time for ESL/ELD progress reports in the agenda for staff meetings.
- Make resources for effective implementation of ESL/ELD programs accessible to staff.
- Allocate budget funds for the purchase of inclusive curriculum resources.
- Consult regularly with board and community resource personnel about additional ways to support and strengthen ESL/ELD programs (Ministry of Education, Ontario, 2001, p. 15).

The initiative is targeted to students with an immigrant background as well as students grown up in a non-English speaking environment. This includes students with Franco-Canadian background as well as students from native Canadian families. Here, a parallel to the situation in Singapore can be observed. Similar to Singapore, also Canada has different language groups living in the country and needed to find ways to accommodate their needs and to find a cooperative way to live in one country. Also parallel to Singapore, Canadian policies appreciate the nondominant languages (other than English) and find it important to protect them. On page 7 of the Ontario resource guide I find: "Research indicates that students benefit academically, socially, and emotionally when they are encouraged to develop and maintain proficiency in their first language while they are learning English. Language skills and conceptual knowledge are readily transferable from one language to another, provided there are no learning exceptionalities. The first language provides a foundation for developing proficiency in additional languages, serves as a basis for emotional development, and provides a vital link with the student's family and cultural background."

A comparison of the percentage of first- and second-generation students who do not speak the language of the test shows that the result is very different for the Canadian states. In Quebec there are 23 % of second-generation immigrant students who do not speak the language of the test at home compared to only 15 % in British Columbia and only 10 % in Ontario. There are two interpretations of this aspect. One is that the populations who immigrated to the different provinces and territories changed significantly in the last 20–30 years. Another interpretation is that although British Columbia's and Ontario's immigrant populations include a high percentage of people who do not speak English after a couple of years, the language spoken at their homes in Canada changes to English upon immigration. In Quebec there is a higher percentage of immigrants that already speak French, but those who do not speak French from the beginning also do not speak French after a couple of years living in Canada. If the latter is the case, policy makers in Quebec might reflect on policies that help immigrants integrate better – especially in terms of language. Language courses for immigrants and especially for children of immigrants might be advisable. Based on these figures – with the caveat that the immigrant population might have changed over time due to the immigration policies but also other factors – one can assume that the immigrant population seems to be well integrated after less than a generation – at least in terms of the language especially in British Columbia and Ontario.

As for the education of the parents, I illustrated in Chap. 4B that similar to Singapore, in the Canadian provinces there are significantly more immigrant students whose parents completed an education of ISCED level 5 or above. In all three provinces there are more than 70 % of first-generation immigrant students whose parents completed an education of ISCED level 5 or above compared to native students with less than 50 % of the parents having completed an education of ISCED level 5 or above. On the other hand, the percentage of students with parents who finished an education of ISCED level 2 or below is quite low in all three groups –

native students, first-, and second-generation immigrant students – with a maximum of 6 % among second-generation immigrant students in Quebec. Obviously, like Singapore, Canada attracts better educated immigrants, which is partly steered by the country's immigration policies as described above.

Another aspect related to the parental background is the parents' socioeconomic status. The number of books at home was used as a predictor for the socioeconomic status. As also found by Brese and Mirazchiyski (2010), the number of books at home is not a good predictor of the socioeconomic status between countries but works quite well within countries. So, although the students in British Columbia and Ontario are among those with the highest average number of books at home, this does not mean that their socioeconomic status is among the highest of the TIMSS participants. But we learned from Chapter 4 that the difference between immigrant students and native students is rather small in all three provinces. Thus we can assume that the socioeconomic status is very similar between native and immigrant students. Statistics Canada found in its research: "In Canada, parental education is less important as a determinant of educational attainment among the children in immigrant families than among those with Canadian-born parents. Less educated immigrant parents are more likely to see their children attain higher levels of education than are their Canadian-born counterparts" (Picot & Hou, 2011). This is an interesting finding that shows that there is more to the impact of immigrant students' parents than their own educational background.

Other literature also identifies parents as one of the factors for the success of immigrant students in Canada. Li (2001) did quantitative research on Chinese parents of immigrant students in Canada. He found that the Chinese parents have very high expectations for the achievement of their children which impacts the children's achievement positively. In fact, on the official Canadian website, the positive results of the immigrant students are explained with their parent's attitude towards their children's education. They state: "Such a fact may not be a surprise to immigrant parents – many of whom chose Canada because of its top-ranked education system. Some experts say that for numerous families, enrolling children in school is one of the first tasks performed upon arrival." And in the following: "'Education is the most important thing for most of these parents, it's why they come here,' says Sharaline Joseph, a settlement worker at the Peel District School Board. Joseph notes that many of her clientele include families with high levels of education, who place extreme value upon learning" (Immigration Canada, 2012).

The attitudes towards school in all three Canadian provinces are more positive in the case of first-generation immigrant students than native students as can be seen in Table 4B.11. Second-generation immigrant students in British Columbia and Quebec also answered more often than their native peers that they like going to school. This can on the one hand be regarded as a positive outcome for immigrant students, but on the other hand, it is also a factor that influences the high achievement of immigrant students in Canada. The same is true for the attitude towards mathematics because in the related question a higher percentage of immigrant students indicated that they like mathematics.

An interesting aspect is the self-rating in mathematics reported in Table 4B.14. In Quebec and Ontario the percentage of students who agree or strongly agree to doing well in mathematics is similar for immigrant and native students. In British Columbia the percentage of students who agree or strongly agree to doing well in mathematics is significantly higher for immigrant students compared to native students. This is a surprising result, considering the countries of origin. In British Columbia more than half of the immigrants come from Asia and the Middle East, with 28 % from Eastern Asia as shown in Table 5B.2. The students in the Eastern Asian countries rate their mathematics abilities in general rather low (see Table 4B.14). This gap between the higher-achieving students in Eastern Asia and their self-confidence is already reported by Lin et al. (2013) or Mullis et al. (2008). In the TIMSS 2003 mathematics report, the authors explicitly find: "It is noteworthy that the four countries with lowest percentages of students in the high self-confidence category – Chinese Taipei, Hong Kong SAR, Japan and Korea – all had high average mathematics achievement. Since all of these countries are Asian Pacific countries, they may share cultural traditions that encourage modest self-confidence" (Mullis, Martin, & Foy, 2005, p. 135). Consequently, the question arises why immigrant students especially in British Columbia, the Canadian province with the highest percentage of immigrants from the Eastern Asian region, have the highest level of self-esteem in comparison to native students.

Summary

The achievement results for students with an immigrant background were very positive in British Columbia and Ontario but somewhat less so in Quebec. This seems to be affected to a large extent by the parental background. The immigration policy in Canada is highly selective and immigrants that have a high education and that are financially affluent are advantaged. The parents of immigrant students also seem to have a positive direct influence on their children's achievement because they put a strong emphasis on high education – this is in particular the case for immigrants from East Asia. Interestingly, this effect seems to be independent of the immigrant's social background.

Furthermore the general positive atmosphere towards immigrants in Canada might have an impact on the positive outcome of immigrant children. This positive attitude towards non-English- or non-French-speaking residents might be caused by the fact that – similar to Singapore – Canada is a multilingual and multicultural society that had to find ways to deal with diversity before the beginning of modern immigration.

The education system of Canada is decentralized and the different provinces have different educational policies. Canada has support programs for students that do not speak the language used for teaching in the schools. These programs target immigrants as well as English-speaking students in Francophone provinces and French-speaking students in Anglophone provinces as well as Aboriginal students.

The English Literacy Development (ELD) programs or English as a second language (ESL) programs give guidance to schools but leave it to schools to implement them in a way customized to the situation.

I conclude that like in Singapore, in the Canadian provinces the selective immigration policies of the country seem to have affected the positive achievement outcome of immigrant students. Canada is a multicultural society which seems to create a fruitful ground for integrating and educating immigrant students. All these aspects can probably not inform other countries' policies to improve immigrant students' achievement, but some of the initiatives might be used as a model for other countries' policies. Like in Singapore, there are support programs for students who have difficulties with the language of instruction which seem to have a positive effect on the students' achievements. It would be valuable if Canada was able to disentangle the effects of the different policy measures.

Chapter 6
Discussion and Conclusion

Abstract The concluding chapter of this publication gives a condensed summary of all the aspects discussed regarding the very diverse group of immigrant student, a topic which has gained increasing relevance in educational research. A high-quality education of immigrant students is not only a question of financial competitiveness but mainly a question of social justice. This publication draws on the most recent research results based on quantitative research. Some of the results of other researchers could be replicated, but some new aspects could be evaluated and may inspire further research. Unlike other research, this publication did not only perform quantitative analyses of large-scale assessment data but also conducted in-depth analyses of countries' policies to help find the policies behind certain research results. This work should not only enhance the knowledge about the situation of immigrant students in various educational systems but also demonstrate how quantitative research using large-scale assessment results can be augmented with policy study results. This summary also points out limitations of this research and suggests starting points for further research.

Keywords Mixed method research • Social justice • Immigrant policies

The publication has addressed one aspect of globalization: the increasing number of immigrant students in various countries. As indicated by the increasing amount of literature, this topic has an increasing relevance in educational research. A high-quality education of immigrant students is not only a question of financial competitiveness but mainly a question of social justice.

This publication draws on the most recent research results based on quantitative research. Some of the results of other researchers could be replicated, but some new aspects could be evaluated and may inspire further research.

Unlike other research, this publication did not only perform quantitative analyses of large-scale assessment data but also conducted in-depth analyses of countries' policies to help find the policies behind certain research results.

This work should not only enhance the knowledge about the situation of immigrant students in various educational systems but also demonstrate how quantitative research using large-scale assessment results can be augmented with policy study results. With this work I hope to have made a contribution to the current discussions in educational research about social justice and immigrant students in particular.

© Springer International Publishing Switzerland 2016

D. Hastedt, *Mathematics Achievement of Immigrant Students*,

DOI 10.1007/978-3-319-29311-0_10

In this chapter I will discuss the results I achieved; I will answer the research questions and evaluate the limitations of the research.

Summary of the Results

In this publication I have recognized children with an immigrant background as one aspect of today's global processes. Immigrant children do have different backgrounds and different reasons why they became immigrants. For example, some are accompanying their parents that are looking for better job opportunities, while others might be refugees. I have noted that there is an increasing number of immigrant students enrolled in several educational systems. In 21 out of 32 countries participating in TIMSS 1999 and TIMSS 2003 grade eight, I found an increasing percentage of immigrant students. This clearly poses a challenge to the educational systems and the countries have applied different measures to react to this.

The overall trend that I noted is that the achievement difference between students with an immigrant background and native students is increasing. I found that while in TIMSS 1995 in 17 out of 37 participating countries, native students outperformed their first-generation immigrant peers in mathematics in grade eight, in TIMSS 2007 in 42 out of 51 participating countries, first-generation immigrant students were achieving statistically significantly below their native peers. In this period the average gap between first-generation immigrant students and native students increased from 14 to 35 score points – a difference that accounts for about 1 year of schooling. For the science achievement I detected similar trends.

For second-generation immigrant students, the results were not that clear. I found some countries with a significant increase in second-generation immigrant students but also a good number of countries with stable statistics and even some countries with a decrease in second-generation immigrant students in the schools in grade eight. In terms of achievement, second-generation immigrant students were lagging behind native students in fewer countries and to a lesser degree than first-generation immigrant students. The results were also quite stable over time.

I then examined some selected background characteristics of the immigrant students and compared them to those of native students. I found that immigrant students – and in particular first-generation immigrant students – were older than their native peers in grade eight. I also examined the hypothesis that immigrant students who arrived in the host country at a younger age perform better than immigrant students who arrived at a later age. This hypothesis could not be confirmed with TIMSS data.

I then investigated how many percent of immigrant students, compared to native students, are boys and girls. Here, I found the most alarming results: in 34 out of 56 educational systems, for example, first-generation immigrant girls are underrepresented in the schools. Out of the 18 countries with the most extreme differences between first-generation immigrant boys and first-generation immigrant girls participating in school education, nine are Islamic countries. The differences come as extreme as in Iran where only 26 % of first-generation immigrant students in the schools are girls against 74 % boys. This means that if I assume a 100 % participation of first-generation immigrant boys and an equal number of boys and girls among the total first-generation immigrant population, probably only every third immigrant girl is enrolled in school. If some first-generation immigrant boys are not enrolled, then the percentage of first-generation immigrant girls enrolled would be even lower than the estimated one out of three.

This issue, however, is not limited to Islamic countries. I found also Asian and Eastern European countries among the participants with low percentages of first-generation immigrant girls in the schools. In Bulgaria and Slovenia, only every second first-generation immigrant girl seems to be enrolled in school, and in Georgia and Lithuania, the percentages are even lower. This is clearly a matter of concern; the background of this result should be evaluated and measures for achieving a high participation of first-generation immigrant girls should be sought. This is not only a question of social justice but simply a matter of human rights.

For second-generation immigrant boys and girls, the picture is much more diverse. I found countries with a lower enrollment of girls but also countries with a lower enrollment of second-generation immigrant boys. But there were substantially fewer cases and the differences were not that pronounced. But also for second-generation immigrant girls, I found educational systems where the percentage of enrolled girls goes as low as 50 %. This is clearly not acceptable in any given country and further research is necessary to fight this injustice.

When I investigated the mathematics achievement of immigrant boys and girls separately, I found that in grade eight the achievement difference for mathematics between immigrant and native students is much more pronounced for boys than for girls. On average, first-generation immigrant girls were lagging behind native girls by 29 score points, but first-generation immigrant boys were lagging behind native boys by 39 score points. For second-generation immigrant students, the results were not that extreme, but also here I found second-generation immigrant girls lagging behind native girls by 10 score points and second-generation immigrant boys being outperformed by 14 points.

Beside these averages I found some countries with quite extreme differences between immigrant boys and immigrant girls in terms of their mathematics achievement. This concerns countries with a higher participation of immigrant boys such as Iran where probably only every third first-generation immigrant girl is enrolled in school but where the ones who actually are enrolled in school are ahead of native girls in terms of their mathematics achievement. First-generation immigrant boys, however, are lagging behind native boys by 47 score points – a difference accounting for about one and a half school years. But also in a country like Turkey where first-generation immigrant girls are at about the same level as native girls with respect to their mathematics achievement, first-generation immigrant boys achieved 52 points below native boys – again a difference accounting for about one and a half school years. The mechanisms behind this also require further research, and measures to support especially immigrant boys in the schools should be considered.

When analyzing whether students are speaking the language of instruction at home, I found that there are more immigrant students than native students speaking a different language at home. This is probably not very surprising, but an interesting aspect is that the difference in mathematics achievement for students who do not speak the language of instruction at home is bigger for native than for immigrant students. Further analysis of the reasons behind this fact may reveal some interesting further results.

I explored two aspects of the students' socioeconomic background. One aspect was the education of the parents and the other the number of books at home – both are relatively good predictors of the socioeconomic status according to the literature. With respect to the parental education, I found quite a number of countries where the educational background of immigrant students was higher than of native students. This applies mostly to first-generation immigrant students but also to second-generation immigrant students. There are countries where immigrant students in this group achieved rather well compared to native students. I suspected that the immigration policies may give preference to potential immigrants who have a better educational background and apply selection criteria for immigration.

In terms of the number of books at home, I found clearly that native students have more books at home than their immigrant peers. We may assume that the number of books at home as a predictor for the socioeconomic status may work differently for native and immigrant students. However, my research has also shown that within each of the groups of immigrant students, the number of books at home is strongly correlated to the students' mathematics achievement.

I evaluated the students' attitudes towards mathematics and towards school in general as well as their self-esteem when it comes to mathematics. The students' attitudes are an important and interesting aspect of education since they can be interpreted on the one hand as an educational outcome and on the other hand as a mediator for learning and consequently as educational achievement. Although I noted some differences for a few countries, in general no difference between native students and immigrant students could be found. It is notable that there was no difference either in the attitude towards mathematics between first-generation immigrant boys and girls. While I found that native boys had more positive attitudes towards mathematics than their native peers in 15 countries, in only three countries first-generation immigrant boys had more positive attitudes towards mathematics than their native peers.

Linking these findings with the achievement differences between immigrant boys and girls, one might suspect that the quite positive attitudes of immigrant girls contributed to the relatively positive outcome in mathematics. (Or one could argue that, considering the bidirectional causality of attitudes and achievement, the relatively positive achievement has contributed to the quite positive attitudes of immigrant girls.) Whichever, trying to influence the immigrant boys' attitudes towards mathematics may have a positive impact on their mathematics achievement, too.

Regarding self-esteem I found in general that native students have a higher self-esteem than immigrant students. And I found similar sets of countries with relatively high self-esteem and relatively high mathematics achievement among immigrant students. An interesting difference was observed in some Arabic countries where the self-esteem of immigrant students was relatively lower than native students' self-esteem although their mathematics achievement was relatively higher. It would be interesting to learn if this controversy is especially pronounced in mathematics or if this is a more general pattern in these countries.

I also analyzed where in the countries immigrant students are living. For a good number of countries, I found more immigrant students in more urban areas. This is true for first-generation immigrant students as well as for second-generation immigrant students. And although the data was sparse, I discovered that in some countries the differences in mathematics achievement between immigrant students and native students were less pronounced in urban areas than in rural areas. I did not find any country with a relative higher achievement of immigrant students in rural areas. This is true for first-generation immigrant students as well as for second-generation immigrant students.

The data available did not allow me to investigate the background of immigrant students who live in rural areas and those who live in urban areas more in-depth. Consequently, I cannot conclude that the populations are comparable. But taking into account these restrictions, I may conclude that some countries seem to be able to accommodate the needs of immigrant students better in more urban areas than in the rural areas. One could investigate if special support programs in rural areas could help immigrant students in these countries.

Different aspects of the school characteristics were analyzed in terms of differences between schools attended by native students and by immigrant students. The subjects of interest here were schools that appear problematic because of low school attendance, schools that are less well resourced, schools with the school climate rated less well – either by the principal or the mathematics teacher – and finally schools where the students feel less save.

In terms of schools with a low student attendance, I found no major differences for native and immigrant students but a slightly higher percentage of first-generation immigrant students in schools with low school attendance. School attendance is associated with mathematics achievement for all students, but in some countries the association tends to be stronger for first-generation immigrant students. Also the resourcing of the schools attended by immigrant students is comparable to the resourcing of the schools attended by native students. However the resourcing of the schools is positively associated with the students' mathematics achievement in some countries.

Furthermore with respect to the school climate, I found it to be positively related to mathematics achievement for native students as well as for immigrant students. This is true for the principals' rating as well as for the teachers' rating. I found native students and immigrant students in general attending schools with similar ratings of the school climate.

The most serious result that was revealed at the school level concerns school safety. In TIMSS 2007, students were asked if something was stolen from them in the last month, if they were hit or hurt by other students, if they were made to do things they didn't want to do by other students, if they were made fun of or called names, and if they were left out of activities by other students. According to my findings in the majority of countries, there is a statistically significantly lower percentage of first-generation immigrant students compared to native students who answered "no" to these five questions. In nearly one third of the participating educational systems, I found the percentage of second-generation immigrant students to be statistically significantly lower than of native students.

This is particularly worrisome as the analysis has shown that for native students as well as for first- and second-generation immigrant students, there is a positive correlation between feeling safe at school and mathematics achievement. Although I cannot conclude a causational relationship between these two parameters, measures should be taken to improve the immigrant students' feeling of safety in school. This is simply a question of humanity and immigrant students' well-being.

In terms of class level factors, I investigated the class size for immigrant and native students, the emphasis of teachers on homework, and the concentration of immigrant students in the classes. With respect to the sizes of the classes attended by immigrant students and native students, I found some differences for a few countries, but in general the class sizes were the same. Surprisingly, I could not find a clear association of class size and students' mathematics achievement. There was also no difference in the emphasis on homework for immigrant and native students even though the emphasis on homework was clearly positively related to the students' mathematics achievement.

More interesting results could be found regarding the concentration of immigrant students in the classes. There are several countries with a high number of immigrant students in some classes. I found 11 countries where the mathematics achievement of first-generation immigrant students was statistically significantly higher when there were only one or two first-generation immigrant students in the class.

As for native students I found also their mathematics achievement to be higher in classes with fewer immigrant students in several countries. Interestingly the association between the number of immigrant students in the class and mathematics achievement is stronger for native students than for immigrant students. But for the countries where the mathematics achievement is higher for immigrant students than native students, I found the mathematics achievement to be positively associated to the number of immigrants in the class.

This suggests that the association that I detected is more an example of the assimilation effect than directly related to the fact that classmates are immigrants. With that conclusion one could argue that any measure taken to improve the achievement of the immigrant students also has a positive effect on the achievement of native students.

Despite some interesting results regarding differences between native students and immigrant students at individual level as well as at school and class level, a few countries always stood out in the results throughout all chapters. Especially Singapore and the Canadian provinces didn't only show positive achievement results for the immigrant students but were also noticeably different in several analyses.

In Ontario and Quebec immigrant students liked going to school more than native students. In Ontario, British Columbia, and Singapore, immigrant students also enjoyed learning mathematics more than their native peers. In Singapore and British Columbia, immigrant students had a higher self-efficacy in mathematics than their native peers. Neither in British Columbia, Ontario, or Quebec nor in Singapore could I find a higher percentage of immigrant students that did not feel safe at school – in British Columbia it was rather the opposite and 53 % of the immigrant students were in the highest category of feeling safe at school compared to only 48 % of native students. In British Columbia, Quebec, and Singapore, a statistically significantly lower percentage of first-generation immigrant students attended classes with a low emphasis on homework. In Singapore and British Columbia, immigrant students in classes with more than two immigrants scored statistically significantly higher than classes with only one or two immigrant students. Also in Singapore and British Columbia as well as in Ontario, I found native students to achieve statistically significantly higher when attending classes with more than two immigrants.

All these aspects made it interesting to further investigate the situation in Canada and Singapore. The last two chapters, therefore, include a more in-depth analysis of the situation of immigrants – and immigrant students in particular – in these countries. The immigration policies and special treatments of immigrant students were investigated.

Singapore has a rather strict and selective immigration policy that grants permission only to well-educated immigrants – in particular when granting permanent residence. An interesting aspect of the Singaporean society is that Singapore is already a multicultural society with three different cultural groups living together. Since conflicts between the different cultures occurred in the past, Singapore already has policies in place to facilitate the communication between citizens with different cultural backgrounds. Since the Singaporean economy requires additional labor force from abroad, policies are in place to increase the acceptance of immigrants by the native citizens.

Singapore introduced policies to help immigrant students get a well-founded start into their school careers. For example, there are rather strict exams in place to grant immigrant students access to the Singaporean educational system which require them to study even before entering into the school system. Moreover governmental programs exist that help students with difficulties in the language of instruction but also in other subjects. As for the Singaporean society in general, there are also policies in place to increase the communication between students and thus the understanding of students with different cultural backgrounds.

For the Canadian provinces I found some similarities. Like Singapore, Canada has very selective immigration policies that give preference to well-educated and more affluent candidates. There is even a trend in immigration policies that puts more emphasis on the education of potential immigrants. Canada can be regarded as a multicultural society – probably not to the same extent as Singapore – but there is a Francophone and an Anglophone part of the population. The original Indian population is respected as a part of the population and their rights are protected. It should also be noted here that the immigrant populations in Canada vary quite substantially between the provinces with respect to their countries of origin.

In Canada there are also language support programs for students who are not fluent in the language of instruction. These can be either students in Francophone schools who are not fluent in French or students in Anglophone schools who are not fluent in English. Since the educational system of Canada is a federal system, the programs differ between provinces and in some provinces even between schools. An interesting aspect of the province of British Columbia is the rather large group of immigrant students with an Asian background. In this group a high value of education seems to be transmitted from the parents to the students and thus a certain pressure of achieving well at school that is independent of their socioeconomic background.

Summarizing the experiences from the analysis but especially from the two country examples, a disappointing result is that countries with higher-achieving immigrant students seem to have in common that their immigration policies are rather restrictive and selective. As a consequence the immigrant students in these countries have a better home background than immigrant students in some other countries.

But it seems also to be important that these countries were already multicultural societies before modern immigration started. Public policies but also policies and traditions in schools paid attention to people with different cultural backgrounds and sought to integrate them into the schools and into the society as a whole.

I also found support programs to assist especially students who are not fluent in the language of instruction. It is probably the mixture of all these aspects that led to the success in teaching immigrant students.

In general, the approach of combining quantitative research and policy analysis has proven to help find and understand differences in immigrant students' education. Quantitative research – especially with cross-sectional data – has the disadvantage that no causal relationships can be found. But finding phenomena in a representative data set can contribute to understanding the existing differences in the population. This can foster the further search for answers to, and connections between, causes and effects.

Analyzing the TIMSS data revealed some interesting aspects and differences in the education of immigrant students. Several aspects have shown to be worthwhile to be investigated further to develop recommendations for policy makers to improve the situation of immigrant students.

Discussion

Today's world is a globalized place which also impacts the educational systems. An increase of immigrant students – as shown in this publication – is only one aspect but one that poses a challenge to the educational systems.

But immigrant students form a quite diverse group. Some come from more affluent homes; others are victims of war and are counted among the most vulnerable
children. What they all have in common is that they deserve a high-quality education and care for their special needs. This is not only a matter of economic advantages but first and foremost a question of social justice.

The analyses of the TIMSS data have revealed some of the differences in the
background characteristics of the immigrant students – especially in terms of their
socioeconomic and parental background but also in terms of how much the language that they speak at home matches the language of instruction at school.

In this publication we have learned that the achievement of immigrant students
is in general lagging behind the achievement of native students. And this achievement gap even seems to increase which shows that most educational systems seem
to be unable to cater adequately for their immigrant student population.

However we have also learned in this publication that the situation of the countries is quite diverse. Not only do they have immigrant student populations with
different backgrounds and of different sizes, but the places where these populations
live in the country – in more rural or more urban areas – vary substantially, which
poses different challenges to the education systems.

Furthermore major differences emerged within the countries that were analyzed
here. Some countries seem to be more successful in achieving better learning opportunities in urban than in rural areas. Then some countries seem to be less successful
in including immigrant girls than immigrant boys in the education system. In other
countries immigrant boys seem to show a bigger achievement gap.

This piece of research highlighted that different countries also seem to have different ways of dealing with the challenge of catering adequately to their immigrant
student population. In some countries immigrant students attend smaller classes or
their teachers seem to have a stronger emphasis on homework. Some countries
appear to be more successful than others.

Singapore and the Canadian provinces of British Columbia and Ontario seem to
be among the more successful ones. But when examining the situation more in
depth, it turned out that a big portion of this "success" seems to stem from the fact
that they all have rather selective immigration policies in place. Consequently, the
aim of this publication to find positive examples from which other countries could
learn was limited and the results somewhat disappointing.

But at least some characteristics and policies could be found which might also
have contributed to the good achievement of immigrant students in these countries.
To mention are, on the one hand, the general positive attitudes towards diversity and
the awareness towards people with different languages and the policies supporting
them. On the other hand we have educational policies to make students aware of
peers with a different cultural background, and educational policies that assist
lower-performing students and students with difficulties in the language of
instruction and that seem to have some positive effect. From the analysis of the
TIMSS data, it became apparent that in these three regions (Singapore, British
Columbia, and Ontario), the teachers put a higher emphasis on homework for immi-

grant students and immigrant students rated their safety at school very high – and both aspects have shown a strong association with the mathematics achievement in most countries.

Although this publication could not find clear answers to the question on how countries should face the challenge of globalization and its impact on the educational systems – especially in terms of the increased immigrant student population – starting points could be determined that deserve some further investigations and that might finally lead to an improved education for immigrant students.

Limitations

There are several aspects that must be mentioned as a restriction for the research presented here. First it should be mentioned that the TIMSS data used in the quantitative analyses has some limitations. TIMSS is a cross-sectional large-scale assessment survey and consequently no causal inferences can be drawn from the data. The analyses could also only use the data available in the TIMSS database. Some aspects that have shown to be very relevant but that the students, teachers, and principals were not asked could not be used. One important aspect that should be mentioned here is the country of origin of the immigrant students.

Another aspect is the available data itself. Although the questions were translated into all target languages and rigorous quality measures were implemented to ensure a high quality of the translations, respondents in different cultures tend to answer questions differently. No cultural invariance of the data from different countries can be assumed, and therefore comparisons of data from respondents with different cultural backgrounds might be subject to artifacts stemming from different response behaviors. Consequently, differences were interpreted very carefully and a major emphasis was put on within country comparisons. In several cases the analysis was carried out for immigrant students and native students separately to ensure that results were not dominated by cultural differences between native and immigrant students.

The analysis of immigrant students' backgrounds showed that these vary quite substantially across countries, especially in the socioeconomic status, the education of the immigrant students' parents, and the language spoken at home. All these are aspects that have shown to impact educational outcomes. It would have been an option to find comparable subpopulations of immigrant students in the different countries for further comparisons but that would have left the sample sizes too small for a meaningful comparison. Therefore, in this publication the student background was analyzed. The interpretation of the results in Chapter 5 was also based on the differences in the students' background.

Finally the publication has a rather narrow focus and analyzed mainly the students' mathematics achievement. Clearly educational outcomes are much more manifold and a high-quality education of immigrant students does not mean that they can compete with native students in terms of their mathematics achievement. And although the science achievement was also considered in chapter 4A and the students' attitudes and self-efficacy, which can also be interpreted as educational outcomes, were analyzed in chapter 4B, it is clear that the main focus is on the students' mathematics achievement. This was necessary to keep the publication manageable in scope. And even though this narrow focus somewhat challenges the aim of finding ways to improve the education of immigrant students, I am willing to argue that if this publication could help improve at least this aspect of the education of immigrant students, it has helped improve the situation for this vulnerable group.

Further Research

Since the analyses and results at hand could not conclude any causal relationships, more research on the effects of the aspects that were found to be different between higher- and lower-achieving immigrant students are needed. Are there causal relationships and if so, what are the mechanisms that, for example, lead to students feeling safe at school? Or does the higher emphasis on homework in Singapore, British Columbia, and Ontario for immigrant students really help their mathematics achievement – and if so, why?

The research in this publication used data of students enrolled in school. I have noted that in some countries certain immigrant students, e.g., immigrant girls, seem not to be enrolled in the education system. It is important to find out more about this group. How many immigrant students are not enrolled in the educational system of the host countries and what are their characteristics? Are they taught outside the schools – maybe privately?

More research projects – especially with more qualitative or longitudinal approaches – should be launched to find ways to better understand the situation of immigrant students and, in the best case, help improve their situation.

References

Alexander, C. (2013). *Speaking note on Canada's immigration policies*. Presented at the Outlook on Immigration and Future Policy, Vancouver: Government of Canada. Retrieved from http://www.cic.gc.ca/english/department/media/speeches/2013/2013-09-26.asp

Alivernini, F., Manganelli, S., & Vinci, E. (2010). An evaluation of factors influencing reading literacy across Italian 4th grade students. *Journal of US-China Education Review, 7*(5), 88–93.

Andersen, S. C., & Thomsen, M. K. (2011). Policy implications of limiting immigrant concentration in Danish public schools. *Skandinavian Political Studies, 34*(1), 27–52. http://doi.org/10.1111/j.1467-9477.2010.00260.x

Baca, R., Bryan, D., & McKinney, M. (1993). The post-middle school careers of Mexican immigrant students: Length of residence, learning English, and high school persistence. *Bilingual Research Journal, 17*(3), 17–39.

Baumann, Z. (2003). *The tourist syndrome*. Retrieved from http://www.google.de/url?sa=t&rct=j&q=&esrc=s&source=web&cd=9&ved=0CHAQFjAI&url=http%3A%2F%2Fred.pucp.edu.pe%2Fwp-content%2Fuploads%2Fbiblioteca%2FZygmunt_Bauman_The_Tourist_Syndrome.pdf&ei=3w4bUsnMIMHq4gTay4HABg&usg=AFQjCNFjnl9jqbKFhihuXEgPWn3omXNgJw&sig2=MQxQOIh983sJ49jnuXZ-DA&bvm=bv.51156542,d.bGE&cad=rja

Beaton, A. E., Martin, M. O., Mullis, I. V. S., Gonzalez, E. J., Smith, T. A., & Kelly, D. L. (1996). *Science achievement in the middle school years: IEA's Third International Mathematics and Science report*. Chestnut Hill, MA: TIMSS & PIRLS International Study Center, Lynch School of Education.

Beaton, A. E., Mullis, I. V. S., Martin, M. O., Gonzalez, E. J., Kelly, D. L., & Smith, T. A. (1996). *Mathematics achievement in the middle school years: IEA's third international mathematics and science report*. Chestnut Hill, MA: TIMSS & PIRLS International Study Center, Lynch School of Education.

Blossfeld, H. P., Bos, W., Lenzen, D., Müller-Böling, D., Prenzel, M., & Wößmann, L. (2008). *Bildungsrisiken und -chancen im Globalisierungsprozess, Jahresgutachten 2008 [Education risks and chances in the process of globalization. Annual report 2008]*. Wiesbaden, Germany: Vereinigung der Bayrischen Wirtschaft e.V., Verlag für Sozialwissenschaften. Retrieved from http://www.aktionsrat-bildung.de/fileadmin/Dokumente/Aktionsrat_Bildung_Jahresgutachten_2008.pdf

Bock, R. D., & Aitken, M. (1981). Marginal maximum likelihood estimation of item parameters: An application of the EM algorithm. *Psychometrika, 46*, 443–459.

Bourdieu, P. (1983). Oekonomisches Kapital, kulturelles Kapital, soziales Kapital. In R. Kreckel (Ed.), *Soziale Ungleichheiten* (pp. 183–198). Göttingen, Germany: Schwartz.

© Springer International Publishing Switzerland 2016
D. Hastedt, *Mathematics Achievement of Immigrant Students*,
DOI 10.1007/978-3-319-29311-0

Bowen, N. K., & Bowen, G. L. (1999). Effects of crime and violence in neighborhoods and schools on the school behavior and performance of adolescents. *Journal of Adolescent Research, 14*(3), 319–342.

Brandt, S. (2008). Estimation of a RASCH model including subdimensions. *IERI Monograph Series, 1*, 51–70.

Brese, F., & Mirazchiyski, P. (2010). *Measuring students' family background in large-scale education studies*. Presented at the 4th IEA International Research Conference in Gothenburg, Sweden, Gothenburg, Sweden. Retrieved from http://www.iea-irc.org/fileadmin/IRC_2010_papers/TIMSS_PIRLS/Brese_Mirazchiyski.pdf

Brophy, J. (2006). *Grade repetition*. Paris, France: The International Institute for Educational Planning (IIEP) & The International Academy of Education (IAE). Retrieved from http://www.unesco.org/iiep/PDF/Edpol6.pdf

Brownlee, P., & Mitchell, C. (Eds.). (1997). Issues paper from Singapore. In *Migration issues in the Asia Pacific*. Wollongong, Australia: APMRN Secretariat Centre for Multicultural Studies - Institute for Social Change & Critical Inquiry; University of Wollongong, Australia. Retrieved from http://unesco.org/most/apmrnw13.htm

Büchel, F., Frick, J. R., Krause, P., & Wagner, G. G. (2001). The impact of poverty on children's school attendance – evidence from West Germany. In K. Vleminckx, & T. M. Smeeding (Eds.), *Child well-being, child poverty and child policy in modern nations* (2nd ed., pp. 151–173). Policy Press Scholarship Online. Retrieved from http://policypress.universitypressscholarship.com/view/10.1332/policypress/9781861342539.001.0001/upso-9781861342539-chapter-7

Buchmann, C., & Parrado, E. A. (2006). Educational achievement of immigrant-origin and native students: A comparative analysis informed by institutional theory. In D. P. Baker & A. W. Wiesemann (Eds.), *The impact of comparative education research on institutional theory* (Vol. 7, pp. 335–366). Bingley, UK: Emerald Group Publishing Limited.

Burbules, N. C., & Torres, C. A. (Eds.). (1999). *Globalization and education: Critical perspectives*. New York, NY: Routledge.

Byrne, B. M. (1996). Academic self-concept: Its structure, measurement, and relation to academic achievement. In B. A. Bracken (Ed.), *Handbook of self-concept: Developmental, social and clinical consideration* (pp. 287–316). Oxford, UK: John Wiley & Sons.

Castles, S. (2009). World population movements, diversity and education. In J. A. Banks (Ed.), *The Routledge international companion to multicultural education* (pp. 49–61). New York, NY/London, UK: Rouledge.

Center for Social and Emotional Education. (2009). *School climate research summary*. New York, NY. Retrieved from http://www.google.de/url?sa=t&rct=j&q=&esrc=s&source=web&cd=3&cad=rja&ved=0CEMQFjAC&url=http%3A%2F%2Fwww.schoolclimate.org%2Fclimate%2Fdocuments%2FschoolClimate-researchSummary.pdf&ei=t5cMUsS6AYmt4ASapIH4DA&usg=AFQjCNF0D9ZIXf50lTDH_UiAV40s8WXEqg&sig2=dVOT4B0axboLauLvlZav8w&bvm=bv.50723672,d.bGE

Chen, X., Rubin, K. H., & Li, D. (1997). Relation between academic achievement and social adjustment: Evidence from Chinese children. *Developmental Psychology, 33*(3), 518–525.

Chong, S. N. Y., & Cheah, H. M. (1997). Demographic trends: Impact on schools. *New Horizons in Education, 58*(1), 1–15.

Cho, R. M. (2011). Are there peer effects associated with having English language learner (ELL) classmates? Evidence from the Early Childhood Longitudinal Study Kindergarten Cohort (ECLS-K). Unpublished working paper. Brown University.

Cliffordson, C., & Gustafsson, J.-E. (2010). *Effects of schooling and age on performance in mathematics and science: A between-grade regression discontinuity design with instrumental variables applied to Swedish TIMSS 1995 data*. Presented at the 4th IEA International Research Conference (IRC), Gothenburg, Sweden.

Cohen, J. (2009). *Jonathan Cohen on school climate: Engaging the whole village, teaching the whole child*. Retrieved from http://www.google.de/url?sa=t&rct=j&q=&esrc=s&source=web

&cd=4&cad=rja&ved=0CEsQFjAD&url=http%3A%2F%2Fwww.edpubs. gov%2Fdocument%2Fed005207w.pdf%3Fck%3D1&ei=cpwMUqDAE-OI4ATYzYGQAw& usg=AFQjCNHZ40wSEgNq1Bh8m6WCo-EYzSZ8yw&sig2=rbv43C8JSjg02G248GgNig&b vm=bv.50723672,d.bGE

Coleman, J. S., Campbell, E. Q., Hobson, C. J., McPartland, F., Mood, A. M., Weinfeld, F. D., et al. (1966). *Equality of educational opportunities*. Washington, D.C.: U.S. Government.

Conger, D. (2010). *Immigrant peers in school and human capital development*. Presented at the "Migration: A World in Motion" multinational conference on migration and migration policy, Maastricht, Netherlands.

Cooper, H., Robinson, J. C., & Patall, E. A. (2006). Does homework improve academic achievement? A synthesis of research, 1987–2003. *Review of Educational Research, 76*(1), 1–62.

Dalit, C. (2011). *Immigrant background peer effects in Italian schools*. Retrieved from http:// www.unito.it/unitoWAR/ShowBinary/FSRepo/D031/Allegati/WP2011Dip/14_WP_Contini. pdf

Department of Statistics, Singapore. (2013). *Yearbook of statistics Singapore 2013*. Singapore, Singapore: Ministry of Trade & Industry.

Di Paolo, A. (2010). *School composition effects in Spain* (Unpublished working paper). Research in Applied Economics Network.

Dirks, G. E. (2006). *Immigration policy*. Retrieved January 24, 2014, from http://www.thecanadi-anencyclopedia.ca/en/article/immigration-policy/

Dronkers, J., & Kalmijin, M. (2013). *Single-parenthood among migrant children: Determinants and consequences for educational performance*. Centre for Research and Analysis of Migration, Department of Economics, University College London. Retrieved from http://www.cream-migration.org/publ_uploads/CDP_09_13.pdf

Dronkers, J., & Kornder, N. (2013). *Can gender differences in the educational performance of 15-year old migrant pupils be explained by the gender equality in the countries of origin and destination?* Maastricht University School of Business and Economics. Retrieved from http:// digitalarchive.maastrichtuniversity.nl/fedora/get/guid:a4e9bf63-aaba-4d9a-82da-9e1f09741f49/ASSET1

Dronkers, J., van der Velden, R., & Dunne, A. (2012). Why are migrant students better off in certain types of educational systems or schools than in others? *European Educational Research Journal, 11*(1), 11–44.

Elley, W. B. (1994). *The IEA study of reading literacy – achievement and instruction in thirty-two school systems*. Oxford, NY: Pergamon Press. Retrieved from http://www.iea.nl/publication_ list.html?&no_cache=1&pubdbSearch=&orderBy=year&order=ASC&filtertypes=&filtervalu e=&resulttype=50&page=2&cHash=0f76a1d253b2d82dc4a62656b552f1a9

Emanuel, R. (2013). *The learning crisis and development*. Washington DC: WorldBank. Retrieved from http://wbgfiles.worldbank.org/video/hdn/ed/symposium2013/Symposium_on_Learning_ Nov7_2013_session1.wmv

Eurydice network. (2009). *Integrating immigrant children into schools in Europe*. Brussels, Belgium: Education, Audiovisual and Culture Executive Agency.

Foy, P., & Olson, J. F. (Eds.). (2009). *TIMSS 2007 international database and user guide (DVD)*. Chestnut Hill, MA: TIMSS & PIRLS International Study Center, Lynch School of Education, Boston College.

Freiberg, H. J. (1998). Measuring school climate: Let me count the ways. *Educational Leadership, 56*(1), 22–26.

Friesen, J., & Krauth, B. (2011). Ethnic enclaves in the classroom. *Labour Economics, 18*(5), 656–663.

Goh, C. T. (2005). *Investment in people pays off for the country: Speech by Senior Minister Goh Chok*. Presented at the Tong at the Jeddah Economic Forum, Saudi Arabia.

Gould, E. D., Lavy, V., & Paserman, M. D. (2005). *Does immigration affect the long-term educational outcomes of natives? Quasi-experimental evidence* (Unpublished working paper). Center for Economic and Policy Research.

Green, A. (1997). *Education, globalization, and the nation state.* New York, NY: St. Martin's Press, Inc.

Haimson, L. (2000). *Smaller is better: First-hand reports of early grade class size reduction in New York City public schools.* Educational Priorities Panel. Retrieved from http://www.class-sizematters.org/wp-content/uploads/2012/11/SmallerIsBetter.pdf

Hansson, A., & Gustafsson, J.-E. (2010). *Measuring invariance of socioeconomic status across migrational background.* University of Gothenburg. Retrieved from http://www.iea.nl/fileadmin/user_upload/IRC/IRC_2010/Papers/IRC2010_Hansson_Gustafsson.pdf

Hattie, J. (2008). *Visible learning: A synthesis of over 800 meta-analyses relating to achievement* (1st ed.). London, UK/New York, NY: Routledge.

Heath, A., & Kilpi-Jakonen, E. (2012). *Immigrant children's age at arrival and assessment results* (OECD Education Working Papers, p. 31). Paris: Organisation for Economic Co-operation and Development. Retrieved from http://www.oecd-ilibrary.org/content/workingpaper/5k993zsz6g7h-en

Hen, N. E. (2010). *Opening address at the 5th teachers' conference 2010.* Suntec City, Singapore. Retrieved from http://www.moe.gov.sg/media/speeches/2010/09/06/5th-teachers-conference-2010.php

Henry, M., Lingard, B., Rizvi, F., & Taylor, S. (2001). *The OECD, globalization and educational policy.* Oxford, NY: Pergamon Press.

Heus, M., Dronkers, J., & Levels, M. (2008). *Immigrant pupils' scientific performance: The influence of educational system features of countries of origin and destination.* Presented at the RC28 Spring Meeting, Florence, Italy.

Heyneman, S. (1993). Educational quality and the crises of educational research. *International Review of Education, 39*(6), 511–517.

Hong, O. S. (2009). Reading abilities of primary one students: Parliamentary replies. Singapore, Singapore: Ministry of Education Singapore. Retrieved from http://www.moe.gov.sg/media/parliamentary-replies/2009/03/reading-abilities-of-primary-o.php

IEA. (2011). *Brief history of IEA: 50 years of educational research.* Retrieved from http://www.iea.nl/brief_history.html

Immigration Canada. (2012). *More immigrant children are entering Canadian school systems.* Retrieved January 24, 2014, from http://www.immigration.ca/en/2012/126-canada-immigration-news-articles/2012/september/399-more-immigrant-children-are-entering-canadian-school-systems.html

Jacobsen, M. H., & Poder, P. (2008). *The sociology of Zygmunt Bauman: Challenges and critique.* Hampshire, UK: Ashgate Publishing Limited.

Jen, T. H., & Chien, C. L. (2008). *The influence of the academic self-concept on academic achievement: From a perspective of Learning motivation.* The Proceedings of IRC 2008. Taipei: IEA, Amsterdam.

Johansone, I. (2010). *Achievement equity by urbanization in Latvia's primary education: Analysis of PIRLS 2006 and TIMSS 2007 data.* Presented at the IEA IRC, Gothenbrg, Sweden: IEA, Amsterdam. Retrieved from http://www.google.de/url?sa=t&rct=j&q=&esrc=s&source=web&cd=1&cad=rja&ved=0CDQQFjAA&url=http%3A%2F%2Fwww.iea.nl%2Ffileadmin%2Fuser_upload%2FIRC%2FIRC_2010%2FPapers%2FIRC2010_Johansone.pdf&ei=_wX1Uv3gBYTDtQb2tIHoDA&usg=AFQjCNHalBwME_qK61fEQ0-l5OzzB0Q1OQ&sig2=1EMU8NnBB9yZZQUHc7_6tQ&bvm=bv.60799247,d.Yms

Kennedy, K. (2012). *Citizenship education and the modern state.* London: Falmer Press.

Kiamanesh, A. R., & Mahdavi-Hezaveh, M. (2008). *Influential factors causing the gender differences in mathematics' achievement scores among Iranian eight graders based on TIMSS 2003 data.* Presented at the IEA IRC, Taipei: IEA, Amsterdam. Retrieved from http://www.iea.nl/fileadmin/user_upload/IRC/IRC_2008/Papers/IRC2008_Kiamanesh_Mahdavi-Hezaveh.pdf

Kozina, A., Rožman, M., Perše, T. V., & Leban, T. R. (2010). *The school climate asa predictor of the achievement in TIMSS advance study: A students', teachers' and principals' perspective.* Presented at the IEA IRC, Gothenbrg, Sweden: IEA, Amsterdam. Retrieved from http://www.iea.nl/fileadmin/user_upload/IRC/IRC_2010/Papers/IRC2010_Kozina_Rozman_etal.pdf

Lauder, H., Brown, P., Dillabough, J.-A., & Halsey, A. H. (Eds.). (2006). *Education, globalization, and social change*. Oxford, UK: Oxford University Press.

Li, J. (2001). Expectations of Chinese immigrant parents for their children's education: The interplay of Chinese tradition and the Canadian context. *Canadian Journal of Education, 26*(4), 477–494.

Lin, S. W., Hung, P. H., & Lin, F. L. (2013). A remedial action based on Taiwanese students' results of TIMSS. Presented at the IEA IRC, Singapore: IEA, Amsterdam.

Lingard, B., & Grek, S. (2007). *The OECD, indicators and PISA: An exploration of events and theoretical perspectives*. University of Edinburgh. Retrieved from http://www.google.de/url?sa=t&rct=j&q=&esrc=s&source=web&cd=1&ved=0CDQQFjAA&url=http%3A%2F%2Fwww.ces.ed.ac.uk%2FPDF%2520Files%2FFabQ_WP2.pdf&ei=I9MdUqmAOoGs7QbUzoFg&usg=AFQjCNHvff_HmNnnOWgfAEKg48S_gD2qkA&sig2=7UGrpPiDBzszJG1DXLp_PA&bvm=bv.51156542,d.bGE&cad=rja

Marsh, H. W., Kong, C. K., & Hau, K. T. (2000). Longitudinal multilevel models of the big-fish-little-pond effect on academic self-concept: Counterbalancing contrast and reflected-glory effects in Hong Kong schools. *Journal of Personality and Social Psychology, 78*(2), 337–349.

Marsh, H. W., & Parker, J. W. (1984). Determinants of student self-concept: Is it better to be a relatively large fish in a small pond even if you don't learn to swim as well? *Journal of Personality and Social Psychology, 47*(1), 213–231. http://doi.org/10.1037/0022-3514.47.1.213

Martin, M. O., & Kelly, D. (Eds.). (1997a). *Third International Mathematics and Science Study technical report Volume II: Implementation and analysis*. Chestnut Hill, MA: TIMSS & PIRLS International Study Center, Lynch School of Education, Boston College.

Martin, M. O., & Kelly, D. (1997b). *TIMSS technical report Volume I: Design and development*. Boston, MA: Boston College.

Martin, M. O., Mullis, I. V. S., Beaton, A. E., Gonzalez, E. J., Smith, T. A., & Kelly, D. L. (1997). *Science achievement in the primary school years: IEA's Third International Mathematics and Science report*. Chestnut Hill, MA: TIMSS & PIRLS International Study Center, Lynch School of Education, Boston College.

Martin, M. O., Mullis, I. V. S., & Chrostowski, S. J. (Eds.). (2004). *TIMSS 2003 technical report*. Chestnut Hill, MA: TIMSS & PIRLS International Study Center, Lynch School of Education, Boston College.

Martin, M. O., Mullis, I. V. S., Foy, P., Arora, A., & Stanco, G. M. (2012). *TIMSS 2011 international results in science*. Chestnut Hill, MA/Amsterdam, Netherlands: TIMSS & PIRLS International Study Center, Lynch School of Education, Boston College/International Association for the Evaluation of Educational Achievement (IEA), IEA Secretariat.

Martin, M. O., Mullis, I. V. S., Gonzalez, E. J., & Chrostowski, S. J. (2004). *TIMSS 2003 international science report: Findings from IEA's Trends in International Mathematics and Science Study at the fourth and eighth grades*. Chestnut Hill, MA: TIMSS & PIRLS International Study Center, Lynch School of Education, Boston College.

Martin, M. O., Mullis, I. V. S., Gonzalez, E. J., Gregory, K. D., Chrostowski, S. J., Garden, R. A., et al. (2000). *TIMSS 1999 international science report: Findings from IEA's repeat of the Third International Mathematics and Science Study at the eighth grade*. Chestnut Hill, MA: TIMSS & PIRLS International Study Center, Lynch School of Education, Boston College.

Martin, M. O., Mullis, I. V. S., Gregory, K. D., Hoyle, C., & Shen, C. (2000). *Effective schools in science and mathematics*. Chestnut Hill, MA: TIMSS & PIRLS International Study Center, Lynch School of Education, Boston College.

Masters, G. N. (1982). A Rasch model for partial credit scoring. *Psychometrika, 47*(2), 149–174.

Mata, M. de. L., Monteiro, V., & Peixoto, F. (2012). Attitudes towards mathematics: Effects of individual, motivational, and social support factors. *Child Development Research, 2012*, 1–10. http://doi.org/10.1155/2012/876028

Meece, J., Glienke, B., & Burg, S. (2006). Gender and motivation. *Journal of School Psychology, 44*(5), 351–373.

Ministry of Education, Ontario. (2001). *English as a second language and English literacy development: A resource guide*. Toronto, ON: Ministry of Education Ontario.

Ministry of Education, Singapore. (2008). *Enhanced learning support programme has benefited pupils*. Retrieved December 10, 2012, from http://www.moe.gov.sg/media/press/2008/01/enhanced-learning-support-prog.php

Ministry of Education, Singapore. (2013a). *Admissions Exercise for International Students (AEIS)*. Retrieved December 10, 2012, from http://www.moe.gov.sg/education/admissions/international-students/admissions-exercise/

Ministry of Education, Singapore. (2013b). *Programmes – learning support*. Retrieved December 10, 2012, from http://www.moe.gov.sg/education/programmes/learning-support/

Mirazchiyski, P. (2013). *Providing school-level reports from international large-scale assessments: Methodological considerations, limitations, and possible solutions* (IEA research report). Amsterdam, The Netherlands: IEA.

Mueller, C. W., & Parcel, T. L. (1981). Measures of socioeconomic status: Alternatives and recommendations. *Child Development, 52*(1), 13–30. http://doi.org/10.2307/1129211

Mullis, I. V. S., Kennedy, A. M., Martin, M. O., & Sainsbury, M. (2006). *PIRLS 2006 assessment framework and specifications* (2nd ed.). Chestnut Hill, MA: TIMSS & PIRLS International Study Center, Lynch School of Education, Boston College.

Mullis, I. V. S., & Martin, M. O. (2008). *TIMSS 2007 encyclopedia*. Chestnut Hill, MA: TIMSS & PIRLS International Study Center, Lynch School of Education, Boston College.

Mullis, I. V. S., Martin, M. O., Beaton, A. E., Gonzalez, E. J., Kelly, D. L., & Smith, T. A. (1997). *Mathematics achievement in the primary school years: IEA's Third International Mathematics and Science report*. Chestnut Hill, MA: TIMSS & PIRLS International Study Center, Lynch School of Education, Boston College.

Mullis, I. V. S., Martin, M. O., Fierros, E. G., Goldberg, A. L., & Stemler, S. E. (2000). *Gender differences in achievement: IEA's Third International Mathematics and Science Study (TIMSS)*. Chestnut Hill, MA: TIMSS & PIRLS International Study Center, Lynch School of Education, Boston College.

Mullis, I. V. S., Martin, M. O., & Foy, P. (2005). *TIMSS 2003 international report on achievement in the mathematics cognitive domains: Findings from a development project*. Chestnut Hill, MA: TIMSS & PIRLS International Study Center, Lynch School of Education, Boston College.

Mullis, I. V. S., Martin, M. O., & Foy, P. (2013). *The impact of reading ability on TIMSS mathematics and science achievement at the fourth grade: An analysis by item reading demands*. Presented at the IEA IRC 2013, Singapore: IEA, Amsterdam. Retrieved from http://www.iea.nl/fileadmin/user_upload/IRC/IRC_2013/Papers/IRC-2013_Mullis_etal.pdf

Mullis, I. V. S., Martin, M. O., Foy, P., Arora, A., & Stanco, G. M. (2012). *TIMSS 2011 international results in mathematics*. Chestnut Hill, MA/Amsterdam, Netherlands: TIMSS & PIRLS International Study Center, Lynch School of Education, Boston College/International Association for the Evaluation of Educational Achievement (IEA), IEA Secretariat.

Mullis, I. V. S., Martin, M. O., Gonzalez, E. J., Gregory, K. D., & Chrostowski, S. J. (2004). *TIMSS 2003 international mathematics report: Findings from IEA's Trends in International Mathematics and Science Study at the fourth and eighth grades*. Chestnut Hill, MA: TIMSS & PIRLS International Study Center, Lynch School of Education, Boston College.

Mullis, I. V. S., Martin, M. O., Gonzalez, E. J., Gregory, K. D., O'Connor, K. M., Chrostowski, S. J., et al. (2000). *TIMSS 1999 international mathematics report: Findings from IEA's repeat of the Third International Mathematics and Science Study at the eighth grade*. Chestnut Hill, MA: TIMSS & PIRLS International Study Center, Lynch School of Education, Boston College.

Mullis, I. V. S., Martin, M. O., Kennedy, A. M., & Foy, P. (2007). *PIRLS 2006 international report: IEA's progress in International Reading Literacy Study in primary school in 40 countries*. Chestnut Hill, MA: TIMSS & PIRLS International Study Center, Lynch School of Education, Boston College.

Mullis, I. V. S., Martin, M. O., Kennedy, A. M., & Foy, P. (2008). *TIMSS 2007 international mathematics report: Findings from IEA's Trends in International Mathematics and Science*

Study at the fourth and eighth grades. Chestnut Hill, MA: TIMSS & PIRLS International Study Center, Lynch School of Education, Boston College.

Mullis, I. V. S., Martin, M. O., Ruddock, G. J., O'Sullivan, C. Y., Arora, A., & Erberber, E. (2005). *TIMSS 2007 assessment frameworks.* Chestnut Hill, MA: TIMSS & PIRLS International Study Center, Lynch School of Education, Boston College.

Mullis, I. V. S., Martin, M. O., Ruddock, G. J., O'Sullivan, C. Y., & Preuschhoff, C. (2009). *TIMSS 2011 assessment frameworks.* Chestnut Hill, MA: TIMSS & PIRLS International Study Center, Lynch School of Education, Boston College.

Mundy, K. (2005). Globalization and educational change: New policy worlds. In N. Bunt, A. Cumming, A. Datnow, K. Leithwood, & D. Livingstone (Eds.), *International handbook of educational policy* (pp. 3–17). Dordrecht, The Netherlands: Springer.

Myers, D., Gao, X., & Emeka, A. (2009). The gradient of immigrant age-at-arrival effects on socioeconomic outcomes in the U.S.A. *International Migration Review, 43*(1), 205–229. http://doi.org/10.1111/j.1747-7379.2008.01153.x

Netten, A. (2010). *Cultural and linguistic diversity in reading literacy achievement: A multilevel approach.* Presented at the 4th IEA international research conference, Gothenburg, Sweden: IEA, Amsterdam. Retrieved from http://www.iea-irc.org/index.php?id=pirls

OECD. (1999). *Measuring student knowledge and skills: A framework for assessment.* Paris, France: OECD Publishing. Retrieved from http://www.oecd.org/edu/school/programmeforinternationalstudentassessmentpisa/33693997.pdf

OECD. (2003a). *Attracting, developing and retaining effective teachers: Country background report for Finland.* Paris, France: Organisation for Economic Co-operation and Development. Retrieved from http://www.google.de/url?sa=t&rct=j&q=&esrc=s&source=web&cd=2&ved=0CDcQFjAB&url=http%3A%2F%2Fwww.oecd.org%2Fedu%2Fschool%2F5328720.pdf&ei=tJf4UvufFcHUtQat14CIBg&usg=AFQjCNFRTK_oRzRpIM5IkGYmwuWkon54Ig&sig2=q5URD2IB5B0Hi1RlUqV7WQ&bvm=bv.60983673,d.Yms

OECD. (2003b). *Student engagement at school: A sense of belonging and participation.* Paris, France: OECD Publishing. Retrieved from http://www.google.de/url?sa=t&rct=j&q=&esrc=s&source=web&cd=3&cad=rja&ved=0CEMQFjAC&url=http%3A%2F%2Fwww.oecd.org%2Fedu%2Fschool%2Fprogrammeforinternationalstudentassessmentpisa%2F33689885.pdf&ei=1LP4UrvkG43dsgafwYDgDQ&usg=AFQjCNFPFwSXtXKS766ZncS6gqV96k1Y3w&sig2=-ijcNe15efMLKcV7SsgOvA&bvm=bv.60983673,d.Yms

OECD. (2005). *Teachers matter: Attracting, developing and retaining effective teachers.* Paris, France: Organisation for Economic Co-operation and Development.

OECD. (2006). *Where immigrant students succeed: A comparative review of performance and engagement in PISA 2003* (1st ed.). Paris, France: OECD Publishing.

OECD. (2010a). *Closing the gap for immigrant students.* Paris, France: OECD Publishing.

OECD. (2010b). *OECD rural policy reviews: Strategies to improve rural service delivery.* Paris, France: OECD Publishing.

OECD. (2010c). *PISA 2009 results: Overcoming social background* (Vol. II). Paris, France: OECD Publishing.

OECD. (2011a). *Against the odds: Disadvantaged students who succeed in school.* Paris, France: OECD Publishing.

OECD. (2011b). *PISA 2009 results: What makes a school successful? Resources, policies and practices.* Paris, France: OECD.

OECD. (2011c). *PISA in focus: When students repeat grades or are transferred out of school: What does it mean for education systems?* OECD Publishing. Retrieved from http://www.google.de/url?sa=t&rct=j&q=&esrc=s&source=web&cd=1&ved=0CC8QFjAA&url=http%3A%2F%2Fwww.oecd.org%2Fpisa%2Fpisaproducts%2Fpisainfocus%2F48363440.pdf&ei=QybhUsq2K8XLswaPqIHwDQ&usg=AFQjCNEHznE9s_Z71qEWbd_tY734ZN_6oA&sig2=xgrv0pbgjCnWj3LNr2P5bg&bvm=bv.59568121,d.Yms

OECD. (2012a). *International migration outlook 2012.* Paris, France: OECD Publishing.

OECD. (2012b). *PISA untapped skills: Realising the potential of immigrant students.* Paris, France: OECD Publishing.

OECD. (2013a). Key skills and economic and social well-being. In *OECD skills outlook 2013* (pp. 223–248). Paris, France: OECD Publishing. Retrieved from http://www.oecd-ilibrary.org/education/oecd-skills-outlook-2013/how-key-information-processing-skills-translate-into-better-economic-and-social-outcomes_9789264204256-10-en

OECD. (2013b). *Settling in: OECD indicators of immigrant integration 2012* (1st ed.). Paris, France: OECD Publishing.

Ohinata, A., & van Ours, J. C. (2011). *How immigrant children affect the academic achievement of native Dutch children* (Discussion Paper Series, Issue 6212). Bonn, Germany: Forschungsinstitut zur Zukunft der Arbeit.

Olson, J. F., Mullis, I. V. S., & Martin, M. O. (2008). *TIMSS 2007 technical report*. Chestnut Hill, MA: TIMSS & PIRLS International Study Center, Lynch School of Education, Boston College.

Organisation for Economic Co-operation and Development. (2005). *Teachers matter: Attracting, developing and retaining effective teachers*. Paris, France: OECD Publishing.

Osborne, J. (2003). Attitudes towards science: a review of the literature and its implications. *International Journal of Science Education, 25*(9), 1049–1079.

Pajares, F., Britner, S. L., & Valiante, G. (2000). Relation between achievement goals and self beliefs of middle school students in writing and science. *Contemporary Educational Psychology, 25*(4), 406–422.

Perše, T. V., Kozina, A., & Leban, T. R. (2008). *Negative school factors and their influence on math and science achievement in TIMSS 2003*. Presented at the IEA IRC, IEA, Amsterdam.

Picot, G., & Hou, F. (2011). *Preparing for success in Canada and the United States: The determinants of educational attainment among the children of immigrants*. Ottawa, Ontario: Statistics Canada. Retrieved from http://www.statcan.gc.ca/pub/81-004-x/200410/7422-eng.htm

Pohl, C. (2006). *Educational achievements of second generation immigrants in Germany*. Presented at the annual meeting of the Scottish Economic Society, Einsteinstrasse, Germany: ifo Institute for Economic Research. Retrieved from https://www.researchgate.net/publication/228647553_Educational_Achievements_of_Second_Generation_Immigrants_in_Germany!

Postlethwaite, T. N., & Ross, K. N. (1992). *Effective schools in reading: Implications for educational planners*. Amsterdam, The Netherlands: International Association for the Evaluation of Educational Achievement (IEA).

Province of British Columbia. (2013). *Going to school: Information for families*. Retrieved January 24, 2014, from http://www.welcomebc.ca/Live/Daily-Life/family-resources/school.aspx

Rasch, G. (1960). *Probabilistic models for some intelligence and attainment tests*. Copenhagen, Denmark: Danish Institute for Educational Research.

Rizvi, F., & Lingard, B. (2006). Globalization and the changing nature of the OECD's educational work. In H. Lauder, P. Brown, J.-A. Dillabough, & A. H. Halsey (Eds.), *Education, globalization, and social change* (pp. 247–260). Oxford, UK: Oxford University Press.

Rizvi, F., & Lingard, B. (2010). *Globalizing education policy*. New York, NY: Routledge.

Robitaille, D. F., & Beaton, A. E. (2002). *Secondary analysis of the TIMSS data*. Dordrecht, The Netherlands: Kluwer Academic Publishers.

Ronnig, M. (2010). *Homework and pupil achievement in Norway evidence from TIMSS*. Statistics Norway. Retrieved from http://brage.bibsys.no/ssb/bitstream/URN:NBN:no-bibsys_brage_27080/1/rapp_201001.pdf

Rubin, D. B. (2009). *Multiple imputation for non-response in surveys*. New York, NY: John Wiley & Sons.

Schwippert, K., Hornberg, S., Freiberg, M., & Stubbe, T. C. (2007). Lesekompetenzen von Kindern mit Migrationshintergrund im internationalen Vergleich [Reading literacy of children with migration background in international comparison]. In W. Bos (Ed.), *IGLU 2006: Lesekompetenzen von Grundschulkindern in Deutschland im internationalen Vergleich* (pp. 249–269). Münster, Germany: Waxmann Verlag.

Settlement.org. (2012). *What kind of ELD support does the school system provide?* Retrieved January 24, 2014, from http://www.settlement.org/sys/faqs_detail.asp?k=ESL_CHILD&faq_id=4001529

Shen, C., & Tam, H. P. (2008). The paradoxical relationship between students achievement and self-perception: a cross-national analysis based on the three waves of TIMSS data. *Educational Research and Evaluation, 14*(1), 87–100.

Shields, K. S., & Behrman, R. E. (2004). Children of immigrant families: Analysis and recommendations. *Children of Immigrant Families, 14*(2), 4–15.

Shumow, L., & Lomax, G. (2001). Predicting perceptions of school safety. *The School Community Journal, 11*(2), 93–112.

Sidhu, R., Ho, K. C., & Yeoh, B. (2011). Emerging education hubs: The case of Singapore. *Higher Education, 61*(1), 23–40. http://doi.org/10.1007/s10734-010-9323-9

Silver, R. E. (2005). The discourse of linguistic capital: Language and economic policy planning in Singapore. *Language Policy, 4*(1), 47–66. http://doi.org/10.1007/s10993-004-6564-4

Simmons, A. B. (2010). *Immigration and Canada: Global and transnational perspectives.* Toronto, Ontario: Canadian Scholars' Press Inc.

Sirin, S. R. (2005). Socioeconomic status and academic achievement: A Meta-Analytic review of research. *Review of Educational Research, 75*(3), 417–453.

Statistics Canada. (2007). *Immigrant population by place of birth, by province and territory (2006 Census).* Retrieved November 23, 2015, from http://www.statcan.gc.ca/tables-tableaux/sum-som/l01/cst01/demo34d-eng.htm

Statistics Canada. (n.d.). *Children of immigrants: How well do they do in school?* Retrieved January 24, 2014, from http://www.statcan.gc.ca/pub/81-004-x/200410/7422-eng.htm

Tappy, N. (2008). *"The secret behind the figures": Evaluation of UPE program in Iganga district.* The Hague, The Netherlands: Institute for Social Studies. Retrieved from http://www.google.com/url?sa=t&rct=j&q=&esrc=s&source=web&cd=3&ved=0CDoQFjAC&url=http%3A%2F%2Fthesis.eur.nl%2Fpub%2F7265%2FTappy%2520Namulondo%2520PPM.pdf&ei=jFBlU8ORGsOgyASp6oLYBQ&usg=AFQjCNGUkn33lDJE_jyzWO4JiK1QEMfBIw&sig2=0p90jUdybjq-_bbpXCDcQw&bvm=bv.65788261,d.aWw&cad=rjt

Trautwein, U., Köller, O., Schmitz, B., & Baumert, J. (2002). Do homework assignments enhance achievement? A multilevel analysis in 7th grade mathematics. *Contemporary Educational Psychology, 27*(1), 26–50.

Trautwein, U. and Köller, O. (2003). The relationship between homework and achievement: still much of a mystery. Educational Psychology Review, 15, 115–145.

UNESCO. (2006). *ISCED 1997: International standard classification of education.* United Nations Educational Scientific and Cultural Organization. Retrieved from http://www.uis.unesco.org/Library/Documents/isced97-en.pdf

UNESCO. (2013a). *Children still battling to go to school* (Policy report No. 10). Paris, France: UNESCO. Retrieved from http://unesdoc.unesco.org/images/0022/002216/221668E.pdf

UNESCO. (2013b). *Toward universal learning: Recommendations from the learning metrics task force.* Montreal, Québec/Washington, DC: UNESCO-UIS/Brookings Institution.

United Nations. (2013). *A new global partnership: Eradicate poverty and transform economies through sustainable development.* New York, NY: United Nations Publications. Retrieved from www.post2015hlp.org/wp-content/uploads/2013/05/UN-Report.pdf

United Nations. (2014). *Global issues – refugees.* Retrieved from https://www.un.org/en/globalis-sues/refugees/

University of Ottawa. (n.d.). *Canada's legal system – sharing of legislative powers in Canada.* Retrieved January 24, 2014, from http://www.slmc.uottawa.ca/?q=laws_canada_legal

von Davier, M., Gonzalez, E., & Mislevy, R. J. (2009). What are plausible values and why are they useful? *IERI Monograph Series, 2*, 9–36.

Vorwurf vom Integrationsrat: Eltern treiben Spaltung an Schulen voran [Accusation by Integration Council: Parents advance divide at schools]. (2012, November 29). *Spiegel Online.* Retrieved from http://www.spiegel.de/schulspiegel/auslaendische-kinder-an-grundschulen-eltern--treiben-spaltung-voran-a-869849.html

Wong, M. (2013). Estimating ethnic preferences using ethnic housing quotas in Singapore. *The Review of Economic Studies (First Published Online).* http://doi.org/10.1093/restud/rdt002

Word, E., Johnston, J., Bain, H. P., Fulton, B. D., Zaharias, J. B., Achilles, C. M., ... Breda, C. (1990). *The State of Tennesse's student/teacher ratio achievement ratio (STAR) project – Final summary report 1985–1990*. Tennessee State Department of Education. Retrieved from http://www.google.de/url?sa=t&rct=j&q=&esrc=s&source=web&cd=4&cad=rja&ved=0CE4 QFjAD&url=http%3A%2F%2Fd64.e2services.net%2Fclass%2FSTARsummary.pdf&ei=Rw UWUsTVHoKVhQe3iYCwCg&usg=AFQjCNFFMONDE2YlNc_t0xxEW9CfUoNJTQ&sig 2=yzZx0Sxm6zq6Mlo765czng&bvm=bv.51156542,d.bGE

World Bank. (2013). *Financing for development post – 2015*. Retrieved from http://www.world-bank.org/content/dam/Worldbank/document/Poverty%20documents/WB-PREM%20 financing-for-development-pub-10-11-13web.pdf

Worswick, C. (2001). *School performance of the children of immigrants in Canada, 1994–98*. Ottawa, Ontario. Retrieved from http://publications.gc.ca/collections/Collection/CS11-0019-178E.pdf

Wößmann, L. (2007). International evidence on expenditure and class size: A review. In *Brookings papers on education policy: 2006–2007*. Retrieved from http://www.google.de/url?sa=t&rct=j &q=&esrc=s&source=web&cd=7&ved=0CG0QFjAG&url=http%3A%2F%2Fwww.brook-ings.edu%2Fgs%2Fbrown%2Fbpepconference%2Fwoessmann_paper.pdf&ei=QwsWUqLrG MjJhAfErYCYCA&usg=AFQjCNEqYYY8_czOa5oSlkPE_t9kU19tgw&sig2=kQ4S-iNvt6u 0y1izcp3YOA&bvm=bv.51156542,d.ZG4

Wößmann, L., & West, M. (2006). Class-size effects in school systems around the world: Evidence from between-grade variation in TIMSS. *European Economic Review, 50*(3), 695–736.

Yeung, R. (2011). *The effect of immigrant composition on students achievement: Evidence from New York City*. Public Administration – Dissertations. Paper 83. http://surface.syr.edu/ppa_etd/83

Yue, C. S. (2001). Singapore: Towards a knowledge-based economy. In S. Masuyama, D. Vandenbrink, & C. S. Yue (Eds.), *Industrial restructuring in East Asia: Towards the 21st century* (pp. 169–207). Singapore, Singapore: Institute of Southeast Asian Studies.